New Developments in
Separation Methods

New Developments
in Separation Methods

Edited by ELI GRUSHKA

Department of Chemistry
State University of New York at Buffalo
Buffalo, New York

*These papers were originally published
in* Separation Science, *Volumes 9 and 10*

MARCEL DEKKER, INC.   New York and Basel

MARCEL DEKKER, INC.
270 Madison Avenue, New York, New York 10016

LIBRARY OF CONGRESS CATALOG CARD NUMBER: 75-37222
ISBN: 0-8247-6411-0

Current printing (last digit):
10 9 8 7 6 5 4 3 2 1

PRINTED IN THE UNITED STATES OF AMERICA

PREFACE

There is hardly a branch of the physical and biological sciences that does not use at least one form of separation method. An organic chemist needs to separate the products of his reaction; a biologist needs to purify some enzymes or proteins; a chemical engineer needs to recover, on a large scale, some petroleum fraction; etc. All too frequently, however, scientists of different discipline tend to (1) overlook separation methods developed by other disciplines and (2) develop their own notations and jargon which makes cross-fertilization a bit more difficult. Thus, as an example, what the biologists call gel filtration the polymer chemists call gel permeation. While it is true that the nature of the column peaking might be different when used by the two groups, the separation mechanism is usually identical. It is also true that frequently a lack of understanding exists between various branches of science. The chemist may have no conception of what the mining engineer is up against and vice versa.

It is clear, then, that it is highly desirable for scientists from various disciplines to get together to exchange opinions, knowledge, interests, and to stimulate one another. It was with this interest in mind that the symposium on New Methods of Separation, which took place during the 167th National American Chemical Society Meeting in Los Angeles, California, April, 1974, was organized. The participants were from various areas of science and the topics included many separation techniques, as can be seen from the papers published here.

It is my belief that a common thread runs through most of the separation techniques. I am rather certain that I am not alone in this belief (see, for example, C. J. King's paper). Hence symposia such as this serve a very useful purpose. That is, to unify the field of separation. The more one knows about different techniques and other approaches to separation, the closer one is to a better understanding of the field as a whole.

It is hoped that the following papers will interest and stimulate all scientists. If this symposium caused some workers to try a separation method previously not used by them, then we have accomplished at least part of our goals.

ELI GRUSHKA

CONTRIBUTORS

E. T. Adams, Jr., Chemistry Department, Texas A&M University, College Station, Texas

R. P. Cahn, Exxon Research and Engineering Co., Linden, New Jersey

Nicholas Catsimpoolas, Biophysics Laboratory, Department of Nutrition and Food Science, Massachusetts Institute of Technology, Cambridge, Massachusetts

W. Charewicz, * University of Kentucky, Lexington, Kentucky

Barnee M. Excott, Chemistry Department, Texas A&M University, College Station, Texas

Will E. Ferguson, Chemistry Department, Texas A&M University, College Station, Texas

R. B. Grieves, University of Kentucky, Lexington, Kentucky

H. W. Hsu, Department of Chemical and Metallurgical Engineering, University of Tennessee, Knoxville, Tennessee

Roy A. Keller, Department of Chemistry, State University of New York, College at Fredonia, Fredonia, New York

C. Judson King, Department of Chemical Engineering, University of California, Berkeley, California

N. N. Li, Exxon Research and Engineering Co., Linden, New Jersey

E. C. Makin, Research Department, Monsanto Polymer and Petrochemicals, St. Louis, Missouri

Michael M. Metro, Department of Chemistry, State University of New York, College at Fredonia, Fredonia, New York

Jerry L. Sarquis, Chemistry Department, Texas A&M University, College Station, Texas

*Institute of Inorganic Chemistry and Metallurgy of Rare Elements, Technical University of Wroclaw, Wroclaw, Poland.

K. Sexton, Department of Chemistry, West Virginia University, Morgantown, West Virginia

P. Somasundaran, Henry Krumb School of Mines, Columbia University, New York

J. H. Strohl, Department of Chemistry, West Virginia University, Morgantown, West Virginia

P. J. W. The, University of Kentucky, Lexington, Kentucky

Carel J. Van Oss, Department of Microbiology, State University of New York at Buffalo, Buffalo, New York

Peter J. Wan, Chemistry Department, Texas A&M University, College Station, Texas

CONTENTS

Selection and Sequencing of Separations

C. JUDSON KING

DEPARTMENT OF CHEMICAL ENGINEERING
UNIVERSITY OF CALIFORNIA
BERKELEY, CALIFORNIA 94720

Abstract

Selection criteria are reviewed for choosing the most promising methods of separation for accomplishing a given separation of chemical components. The degree of separation attainable is governed by the separation factor, which in turn is related to molecular or ionic properties. Various separation processes are contrasted with regard to general desirablilty of using them, both on an industrial scale and on a laboratory scale. Heuristic rules for synthesizing sequences of separations to create multiple products are also considered. Criteria which have proven useful for generating separation sequences on an industrial processing scale are contrasted with those which may be useful for choosing sequences of separation steps in the laboratory.

INTRODUCTION

The problems of selecting a method of separation to accomplish a given goal and of selecting the optimal ordering of separations are both common to industrial chemical processing and to laboratory chemical analyses. The purpose of this short review is to consider methods which are useful for picking separation methods and sequences, and to contrast criteria suitable for industrial processing with those applying to the laboratory scale.

Common Features of Separation Processes

Common features of separation processes have been discussed by the author (1) as well as by others. Any separation must be based upon a

1

separation factor, akin to the relative volatility for a distillation, which denotes the extent to which the products from the separation differ in composition. The separation factor α_{ij} is defined by

$$\alpha_{ij} = \frac{x_{i1}x_{j2}}{x_{i2}x_{j1}} \tag{1}$$

where subscripts i and j refer to the components being separated and subscripts 1 and 2 refer to two different products. x is mole fraction or any other suitable composition parameter.

The separation is generally effected by means of a *separating agent*, which is a stream of matter or energy fed to the process to cause the separation to occur. Since separation is the opposite of mixing and since mixing is an irreversible process, purposeful design and a certain minimum energy consumption are required for a separation process. This energy is usually provided by means of the separating agent.

Categorizations of Separation Processes (*1*)

Some separations simply segregate two phases already present in a mixture on the microscale into two collected products, one of each phase. These are known as *mechanical* separation processes. We shall be concerned here with separations of homogeneous mixtures, which may include a mechanical separation of phases as a final step.

Separation processes may be categorized as being either *equilibration* processes or *rate-governed*. An equilibration process is based upon the approach of two immiscible phases toward equilibrium compositions which differ from one another. Rate-governed processes are based upon different rates of transport of different species through some medium, often a barrier between feed and product.

Separation processes can also be categorized according to the type of separating agent, with *mass separating agent* processes (such as absorption and extraction), which utilize solvents or other streams of matter, being distinguished from *energy separating agent* processes (such as distillation and crystallization), which supply the separating agent as heat, cooling, or compression work.

Differences between Industrial Processes and Laboratory Analyses

The choice and design of industrial separation processes are dominated by economics, whereas the choice and design of separations used for analytical purposes in the laboratory are more influenced by convenience (short time, little effort), flexibility, and attainable completeness of separation. Industrial processing separations tend to be continuous flow processes, while those in the analytical laboratory are for the most part batch processes. Industrial separations most often involve large quantities of material, whereas analytical separations involve small quantities. In fact, the ability to handle extremely small quantities is often a prime virtue in analytical methods.

For purposes of chemical analysis it is not necessary that the various species analyzed be separated from one another into products of different composition. Techniques such as spectrometry and NMR allow quantitative identification of relative amounts of different species without formal separation. Such approaches depend upon the degree of resolution between components rather than the degree of separation. Methods of chemical analysis can often be destructive, in that the chemical nature of the species being separated or identified can be changed or lost, as in pyrolysis. On the other hand, the usual purpose of industrial separations is to isolate individual components as salable products, as reactor feeds, or as recycle streams. In all these situations the chemical nature of the different components should not be changed.

SELECTION OF SEPARATION PROCESSES

Separation Factor

The separation factor results from differences in appropriate physical properties, such as vapor pressure and solubility. These differences result in turn from differences in molecular properties, a list of which is given in Table 1. In order for separation to be possible the separation factor must be different from unity, the more removed the better.

Different separation processes involve separation factors which are dependent to different extents upon the different molecular properties listed in Table 1 (*1*, Table 14–1). For example, separation by distillation depends upon differences in volatility, which in turn come from differences in molecular weight and in intermolecular forces, reflected by dipole moment and polarizability. Crystallization, on the other hand, involves the ability of molecules to fit into an ordered solid-phase structure and thereby relies more upon differences in molecular volume and molecular shape for determining separation factors. Electrophoresis requires differences in molecular charge, ultracentrifugation requires differences in

TABLE 1

Molecular Properties upon Which Separations May Be Based

Molecular weight
Molecular volume
Molecular shape
Dipole moment
Polarizability
Electrical charge
Chemical reactivity

molecular weight, and ion exchange requires differences in chemical reactivity with the exchange resin. Solvent extraction provides separation factors through differences in activity coefficients of the components within the solvent and feed phases; these differences in turn usually result from differences in dipole moment and polarizability. Foam or bubble fractionation requires differences in surface activity which in turn result from differences in intermolecular forces.

Since separation factors for different processes depend to different extents upon differences in different molecular properties, one can judge a priori which separation processes are likely to give the most favorable separation factors by evaluating the molecular properties in which the species to be separated differ to the greatest extent.

Nature of the Process

There are certain inherent features of different separation processes which favor or disfavor use of the process for industrial-scale processes and for laboratory analytical separations. In fact, it is striking to observe that only rarely is the same separation method for a given component mixture used in a chemical process and for laboratory analyses.

Certain processes are inherently better suited to large-scale or to small-scale operation. For example, gas–liquid chromatography sees widespread use for analyses because of its flexibility and ability to handle multicomponent systems in a single operation, but it has for the most part resisted efforts to achieve economical operation upon scale-up. Similarly, electromagnetic separation with the mass spectrometer is one of the most successful separation methods for analysis, but is generally not suitable for separations on an industrial scale. An exception is the use of the Calutron for production-scale separation of radioisotopes; however, that usage is a result of a unique historical situation (2).

Some separations are better adapted to batch operation than to fully continuous operation. A good example again is chromatography of any sort, and this is another reason why chromatography sees much more use in the laboratory than on a production scale.

Distillation, on the other hand, is a process which sees more use for industrial separations than for chemical analyses. Distillation is well-suited to large-scale, continuous operation and is usually a relatively inexpensive separation with considerable economies of scale. Batch distillation is used for analyses in some cases, especially for complex petroleum mixtures, but suffers from relatively incomplete separations resulting from the lack of a stripping section.

For industrial separations, energy-separating-agent processes tend to be relatively favored over mass-separating-agent processes because of the extra quantities of material to be handled in the case of mass separating agents and because of the expense associated with subsequent separation and regeneration of the mass separating agent from one of the products. These considerations are much less important for laboratory analyses; in fact, it is usually not necessary to regenerate a mass separating agent when one is used. For this reason, precipitation sees much more widespread use as a separation process for laboratory analyses than for industrial separations. Precipitation is the basis of the classical qualitative analysis schemes and of all gravimetric analysis. When used on a large scale, however, precipitation suffers from the facts that the material being separated is chemically converted and that there is a net consumption of a precipitant which will often be a major process expense. Furthermore, there must usually be an excess of precipitant added, and this will appear in a product, often contributing further separation problems.

Another characteristic of gas–liquid chromatography is the considerable dilution of the species being separated with the carrier gas, which is a mass separating agent. This is of little concern for analyses but represents a large economic burden if the process is to be used where recovery of the components in pure form is needed.

Requirement of extreme processing conditions can be a deterrent on both the laboratory and industrial scales. Thus one does not ordinarily contemplate separating a mixture of salts by distillation in either case. Extreme conditions of temperature or pressure lead to expense and can lead to product damage; both these aspects tend to discourage the use of extreme conditions on an industrial scale even more than for laboratory analyses.

Use of large-scale separations involving a solid phase poses problems.

Usually truly continuous-flow operation for both phases is not possible in an economic way, and the approach instead is to use a fixed-bed process, involving intermittent operation and regeneration of solid material in the fixed bed. Handling and transport of solids and slurries also poses problems. These difficulties are of much less concern for laboratory analyses, and therefore separations with solid phases (e.g., adsorption chromatography and precipitation) are much more common.

Some separation methods lend themselves much more readily to multiple staging than do others Rate-governed separation processes, such as reverse osmosis, are difficult to stage because of the need for using a separate device for each stage and supplying separating agent in the form of energy to each stage. Thus rate-governed processes are relatively more preferred when only one or a few stages are necessary. Since separations for chemical analysis often require very high degrees of resolution of components and hence many stages, rate-governed processes see less utilization for analyses.

Two other pertinent selection criteria apply both to industrial-scale separations and to laboratory analyses. One of these is a desire to choose separations which are capable of providing the desired grouping of components into products. As an example, we may want to separate a mixture of aromatic and paraffinic hydrocarbons of approximately the same molecular weight range. Boiling points of the aromatics and paraffins will overlap, so that it will not be possible to achieve the desired component grouping into the products of a single distillation. On the other hand, solvent extraction with a relatively polar solvent will often be able to remove all the aromatics preferentially to the paraffins in a single step. No matter what the scale of operation, additional separation steps beyond the minimum necessary will incur a penalty in expense and/or effort.

Another criterion involves experience. For good reason the industrial designer and the laboratory analyst stick with tried and true processes, since development of a new separation method for each design or each analysis is inefficient.

An Example—Separation of Xylene Isomers (*1*)

Different xylene isomers are used for different purposes in the chemical industry. *p*-Xylene is a raw material for the manufacture of terephthalic acid, an intermediate for polyesters. *o*-Xylene sees use for the production of phthalic anhydride and thereby plasticizer compounds and other derivatives. *m*-Xylene is largely recycled and isomerized into the para and ortho compounds, although there are some developmental uses of the meta isomer. Thus separations of the isomers are important to the petrochemical industry, as well as being important for analyses.

Referring to Table 1, the xylene isomers do not differ in molecular weight, molecular volume, polarizability, or electrical charge. They differ slightly in dipole moment, since the ortho isomer and to a lesser extent the meta isomer do possess nonzero moments. They differ significantly in molecular shape and consequently differ in chemical reactivity with some reactants where steric effects are important. Molecular shape has the greatest influence upon separations involving a solid phase, such as crystallization and adsorption. Correspondingly, one finds that the most established process for industrial separation of p-xylene from m-xylene is crystallization, which is based upon a substantial difference in freezing points—$+56$ and $-55°F$ for p-xylene and m-xylene, respectively. Another successful new process is based upon adsorption.

The slight dipole moment (0.62×10^{-18} esu) of o-xylene gives a boiling point ($291.2°F$) which is significantly greater than those of m-xylene ($282.7°F$) and p-xylene ($281.3°F$). Even though the resultant relative volatility of the other isomers to the ortho isomer is only 1.16, distillation is commonly used for the recovery of o-xylene industrially. A reflux ratio of 15:1 and 100 or more plates are required. This is an example of the very great attractiveness of distillation compared to other separation processes industrially, since much better separation factors are available with several other processes. The very small relative volatility (1.02) between m-xylene and p-xylene renders distillation uneconomical for their separation industrially. Again, because of the small difference in dipole moments, the addition of polar solvents for azeotropic distillation of the meta and para isomers has not been successful, with the best solvent among 40 tested merely improving the relative volatility to 1.03.

It is interesting to observe that none of these approaches are used for laboratory analyses of xylene isomers. Instead, one frequently takes advantage of the fact that the original chemical nature of the isomers need not be preserved. Thus the isomer mixture is often converted by $KMnO_4$ oxidation to the dicarboxylic acids, which have quite different melting points from one another. The same is true of the nitro derivatives. The other common laboratory procedure, of course, is gas–liquid chromatography, which takes advantage of the incentives which have already been noted for chromatography. Even though quite low relative volatilities between the isomers are achieved with most solvents and column substrates used, the ability of GLC to provide the effects of very large numbers of stages in improving the separation allows a quantitative resolution of the isomers to be made with sufficiently good chromatographic systems.

SEQUENCING OF SEPARATIONS

When several products containing specified groupings of components are to be obtained from a mixture, the number of possible sequences of separators becomes quite large. Table 2 illustrates the number of possible sequences (S_R) for separating a mixture of R components into single-component products. It is assumed in Table 2 that a single method of separation is available, in which case the series

$$S_R = \sum_{j=1}^{R-1} S_{R-1} S_j \tag{2}$$

is obeyed. When several different kinds of separators may be considered for use, the numbers become even larger. For example, if N different separators may be considered at each constituent step in the sequences, and if the ordering of separation factors is the same for all separation methods, then the number of different sequences grows by a factor of N^{R-1}, since each of the $R - 1$ separators in each scheme could be replaced by any of the other separation methods.

The situation becomes more complex in the case where the different products should contain specified mixtures of components and/or the separation factors for different methods of separation have different orderings. Because of the orderings of separation factors, only certain combinations of separators will have the capability of producing the desired components groupings in products without splitting sets of components that may remain together. If a sequence is used which separates

TABLE 2

Number of Separator Sequences

No. of components, R	No. of separators, $R - 1$	No. of sequences, S_R
2	1	1
3	2	2
4	3	5
5	4	14
6	5	42
8	7	429
10	9	4862

components unnecessarily, then more than the minimum number of separators will be employed. In general, using extra separations will lead to greater expense on the industrial scale, and almost certainly will lead to greater effort on the laboratory analytical scale. Thompson and King (3) describe a systematic procedure which may be implemented by computer if desired, and which has the capabiity of identifying those sequences which can make the desired separations into products with the minimum number of separators. Two matrices, a pair compatibility matrix and a product separability matrix, are employed. If no sequence can lead to the desired component groupings in products without splitting a product, then the procedure identifies those sequences which involve the least number of product splits.

The separator selection problem has a tree structure in that there are subproblems corresponding to the separation of lesser numbers of components into products. After the initial separation steps in a sequence have been fixed, certain of these subproblems remain. The same subproblems can show up following various different initial steps. If a single cost can be attached to the use of a particular separator for separating a particular mixture into a particular set of products, no matter what separations come before or after, then dynamic programming may be used as a very efficient procedure for determining the optimal separation sequence (4). The assumptions necessary for this approach can be significantly limiting for industrial separation problems, but are probably better fulfilled for separation sequences used for laboratory analyses, such as classical qualitative and quantitative analysis schemes.

For separation problems where a large number of alternative sequences are possible, it is usually true that the difference in cost or amount of laboratory effort does not differ much from the true optimal sequence for the next several sequences closest in attractiveness to the optimal sequence. In such cases the incentive for finding the single optimal sequence diminishes, and it is probably satisfactory just to find a sequence which is close to optimal, or to find a number of near-optimal sequences and make the final judgment among them on the basis of other qualitative factors. In these instances heuristic rules are useful for synthesizing separation sequences.

A listing of heuristic rules which have been found useful for the selection of distillation sequences on an industrial scale is given in Table 3 (1). The desirability of using the minimum number of separators follows from the discussion above. Performing difficult separations, in terms of either separation factor or product purities, last is desirable from the standpoint of

TABLE 3

Heuristic Rules for Sequencing Distillation Columns

Minimum number of separators (or products)
Perform tight (low separation factor) separations last,
  as binaries
Make high purity products last
Take one component at a time overhead
Use descending pressure level of columns
Favor approximately equal product flows

minimizing the amount of additional components present, since difficult separations require many stages and/or much separating agent. Taking one component at a time overhead in distillation sequences leads to the minimum vapor generation rate, if the relative volatilities for all separations are about the same, and also leads to the least number of high-pressure columns if pressures above atmospheric are required. On the other hand, taking approximately equal amounts of both products is desirable from the standpoint of properly utilizing the balance between the reflux rate in the rectifying section and the vapor boil-up rate in the stripping section of a ·distillation column. These last two heuristics often conflict. The use of heuristic rules for the assembly of separation sequences has been explored by Thompson and King (3). One finding is that an effective heuristic is to use next the cheapest separator from among that group of potential next-separators which have the capability of leading to the minimum number of products. This heuristic combines several of those from Table 3.

The synthesis of industrial separation schemes has been reviewed recently by Hendry, Rudd, and Seader (5).

The heuristics to use for laboratory analytical separation sequences would be largely different. The incentive for utilizing the minimum total number of separations should remain valid. For any analysis one would seek separations which can give a high resolution between components, and which can create the desired split between components without great effort or extreme conditions. For qualitative analysis it should be desirable to seek sequences which have the potential of confirming the absence of large groups of components and of dividing the possible components into groups each containing a substantial number of components. One can see this factor at play in the design of the standard qualitative analysis schemes. For example, if a precipitate is not obtained for any major group, then all those components are known to be absent. In the case of quantitative analyses of components that are already known to be present, there is a greater incentive for separating one component at a time and doing it in such a way (e.g., precipitation) so as to allow for the quantitative determination of that component as it is separated. Finally, perhaps the most important point to be made regarding separations for analyses is that a single-operation multicomponent separation—such as chromatography —has by far the greatest efficiency and is satisfactory for the purpose desired.

REFERENCES

1. C. J. King, *Separation Processes,* McGraw-Hill, New York, 1971, Chaps. 1, 13, and 14.
2. L. O. Love, *Science, 182,* 343 (1973).
3. R. W. Thompson and C. J. King, *Amer. Inst. Chem. Eng. J., 18,* 941 (1972).
4. J. E. Hendry and R. R. Hughes, *Chem. Eng. Progr., 68*(6), 71 (1972).
5. J. E. Hendry, D. F. Rudd, and J. D. Seader, *Amer. Inst. Chem. Eng. J., 19,* 1 (1973).

Separation of Phenol from
Waste Water by the Liquid Membrane Technique

R. P. CAHN and N. N. LI

EXXON RESEARCH AND ENGINEERING CO
LINDEN, NEW JERSEY 07036

Abstract

The removal of phenol and other weakly ionized acids and bases from waste water is described when using the liquid membrane emulsion technique. Mathematical relationships are derived for the theoretical distribution and for the rate of permeation of phenol into the emulsion.

GENERAL DESCRIPTION OF LIQUID
MEMBRANE SEPARATION PROCESSES

Liquid membranes, in general, are formed by first making an emulsion of two immiscible phases and then dispersing the emulsion in a third phase (continuous phase). The liquid membrane phase refers to the phase in between the encapsulated phase in the emulsion and the continuous phase. Usually the encapsulated phase and the continuous phase are miscible, but they are not miscible with the membrane phase (1). The emulsion can be either oil-in-water or water-in-oil. By definition the liquid membrane then will be of the water type in the former case and of the oil type in the latter case. The liquid membrane phase usually contains surfactants, additives, and a base material which is a solvent for all the other ingredients. The surfactants and additives are used to control the stability, permeability, and selectivity of the membrane. For specific applications, liquid membranes can be tailor-made.

When the emulsion is dispersed by agitation in a continuous phase (the third phase), many small globules of emulsion are formed. These globules are stable and do not disintegrate, as would be expected, when the system is agitated. Their size depends strongly on the nature and concentration of the surfactants in the emulsion, the emulsion viscosity, and the mode and intensity of mixing. In most of our laboratory runs, the size was controlled in the range of 1 to 2 mm diameter. Each emulsion globule contains many tiny encapsulated droplets with a typical size of 1 μ in diameter. A large number of globules of emulsion can easily be formed to produce a correspondingly large membrane surface area for rapid mass transfer from either the continuous phase to the encapsulated phase or vice versa. In addition, "facilitated transport" mechanism may be used to enhance the mass transfer. This is usually achieved by incorporating a reagent in the encapsulated phase which can react with the permeating compound from the continuous aqueous phase (2, 3), or by incorporating a compound, such as a complexing agent, in the membrane phase to increase the solubility of the permeating compound in the membrane, and therefore the permeation rate of this compound through the membrane (4).

After a desired degree of separation has been achieved, mixing is stopped and the globules of emulsion quickly coalesce and form a layer of emulsion. The emulsion phase can be lighter or heavier than the continuous phase and can be easily separated from the continuous phase. The contacting operation can be either batchwise or continuous, cocurrent or countercurrent.

EXPERIMENTAL

Equipment and Procedure

The water treatment data reported previously (3) and in this paper were all obtained from small-scale laboratory experiments. The equipment employed to carry out the separation was quite simple. It involves mainly a mixer with a stirrer.

In a typical laboratory run, the aqueous reagent solution to be encapsulated by liquid membranes is poured at a rate of about 10 cc/min into 200 cc of the hydrocarbon-surfactant solution at a mixing rate of about 1200 rpm. The final weight ratio of the aqueous reagent solution to the membrane-forming solution is usually 1:1. The total time for emulsifying the entire aqueous solution in the membrane-forming solution is about 15 min. The resulting emulsion droplets usually have a diameter of 10^{-3} to 10^{-4} cm. The waste water, containing the contaminant, such as phenol or ammonia, is contacted with the above emulsion in a mixer. Usually good dispersion of the emulsion in the water to be treated is maintained by the use of agitation with a mixing speed of 100 to 200 rpm for a period of about 5 to 20 min. During a run, agitation may be stopped from time to time so that feed sample can be taken for measurement of pH and contaminants concentrations. Waste water to emulsion weight ratio usually varies from 1 to 10, depending on the reagent concentration used which is usually from 0.1 to 20 wt%.

Materials

The reagents sodium hydroxide and sulfuric acid were of c.p. grade. The oil used as the solvent in the membrane phase was dewaxed Solvent 100 Neutral, which is a middle distillate having an average molecular weight of 386.5 and a density of 0.836 g/cc measured at 25°C. The surfactant used was Span-80, which is sorbitan monooleate, manufactured by Atlas Chemical Co.

SEPARATION MECHANISM AND RESULTS

Facilitated Transport

When a water-in-oil emulsion is dispersed in an aqueous phase, the system will consist of individual stable emulsion globules floating in the water phase as shown in Fig. 1. Contaminants dissolved in the continuous aqueous phase will diffuse through the oil membrane into the small drop-

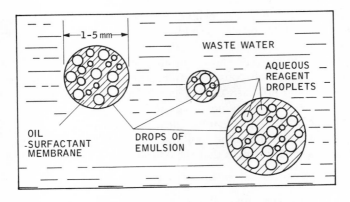

FIG. 1. Water treatment by LM-emulsion.

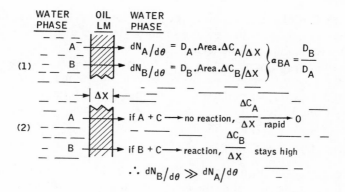

FIG. 2. Two LM separation mechanisms.

lets on the inside of the emulsion as long as there is a driving force for this diffusion, and provided that there is some solubility of the contaminating species in the liquid membrane phase.

A simplified picture of what takes place is shown in Fig. 2. On the left is the waste water stream, or the continuous aqueous phase which contains contaminants A and B. Separating the waste water phase from the aqueous droplet phase on the right, and forming a coherent film, is the liquid membrane oil phase. The upper example (1) shows one mechanism which can accomplish separation of components A and B dissolved in the aqueous phase on the left. This mechanism relies on differential rate of permeation through the membrane. If the permeability of B > permeability of A, B will tend to concentrate in the permeate on the right, while a solution rich in A will stay behind. The reason for the higher permeability of B through the oil may be a higher diffusion coefficient, or a higher solubility of B in the oil phase, or both.

The lower example (2) shows another mechanism, namely how a chemical reaction which consumes B but not A can make this a very selective separation process. As long as reagent C and the product of the reaction between B + C are not permeable through the membrane. this system can be used to accomplish nearly complete as well as highly selective removal of B from the solution on the left.

Essentially what happens is that the concentration (c) driving force

$$\Delta c = c_{left} - c_{right} \tag{1}$$

for compound B in the rate equation

$$dN/d\theta = (D)(area)(\Delta c/\Delta x) \tag{2}$$

(where $dN'/d\theta$ is the quantity of material permeating across a given area of membrane/unit time, Δc is the concentration difference of the permeating species on either side of the membrane, Δx is the membrane thickness through which permeation takes place, and D is the diffusion coefficient of permeating species through the membrane) is maintained high by keeping c_{right} close to zero by the reaction which consumes B.

In short, while separation mechanism (1) is due to differential rates of permeation of the several constituents of a mixture through the liquid membrane, mechanism (2) relies on chemical reaction to affect the driving force of the material to be transferred. In essence, it is providing a sink for the permeating material by reacting it chemically inside the emulsion.

A typical example of such a system is the caustic-in-oil emulsion shown in Fig. 3 which can be used effectively to remove small amounts of phenol from a waste water stream. This example will be discussed in greater detail below. However, what happens, in brief, is that phenol, being somewhat oil soluble, will permeate readily from the outside water phase

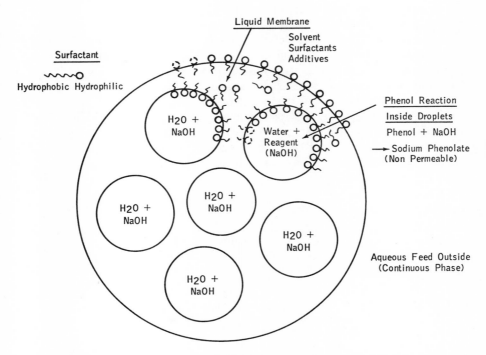

Fig. 3. Schematic diagram of liquid membrane system for phenol removal.

through the oil membrane into the internal aqueous caustic droplets. Here the phenol will be neutralized by the caustic and tied up as sodium phenate which is insoluble in oil and, consequently, cannot diffuse back out again.

Typical examples for the removal of dissolved constituents in waste water by means of the liquid membrane technique are weak acids or bases which are soluble in the liquid membrane. They can therefore diffuse through the membrane in their undissociated form and will be neutralized by the solution of a strong base or acid, respectively, which has been dispersed in the emulsion formulation. Since neither the strong acid or base, nor the resultant ionized salt, are oil soluble, they will remain encapsulated inside the emulsion drops.

Table 1 lists a few examples of waste water constituents which we have removed successfully by means of liquid membrane emulsions.

In all cases the mechanism is the same. The impurity in the waste water, just as with phenol, will permeate across the oil membrane in its undissociated form and will be neutralized by the strong caustic or acid inside the aqueous droplets within the emulsion, preventing its escape back out into the waste water.

Before going into more detail on the mechanism, it will be useful to outline briefly some typical installations which can be considered for the removal of contaminants using the liquid membrane technique.

Figure 4 shows a single-stage installation together with emulsion make-up, recycle, and disposal. If more than one stage is required, as will be discussed later, cocurrent mixers, countercurrent mixer-settler stages, or a mechanically agitated plate tower may be used (Fig. 5). It is unlikely that a simple plate tower used in liquid/liquid extraction will give sufficient agitation to provide the necessary contacting between the two phases.

TABLE 1

Removal of Weak Acids and Bases

Acidic Materials Removed by "Caustic" LM Emulsion
Phenol
H_2S
HCN
Acetic acid
Other organic acids
Basic Contaminants Removed by "Acid" LM Emulsion
NH_3
Amines

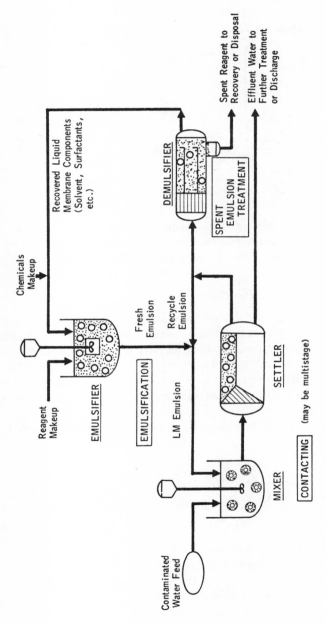

FIG. 4. Schematic diagram of conceptual LM water treating process.

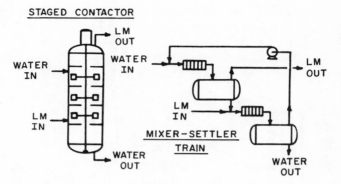

FIG. 5. Multistage countercurrent contactors.

As mentioned previously, the operation may be carried out either batch-wise or in a continuous fashion. Flow of waste water and emulsion phases may be cocurrent or countercurrent.

Equilibrium Distribution Curve

By encapsulating into the emulsion an aqueous solution of either a strong base or acid, it is possible to remove weakly acidic or basic constituents from waste water, provided they are reasonably soluble in the oil phase and also that they have a relatively low dissociation constant.

Phenol is a good example of how to calculate what degree of removal can be achieved in a liquid membrane treater, using the phenol dissociation constant as a guide.

The dissociation constant of phenol in water at 25°C is 1.28×10^{-10}, or

$$\frac{[PhO^-][H^+]}{[PhOH]} = 1.28 \times 10^{-10} \tag{3}$$

Also, at 25°C the ionization product of water is close to 10^{-14}, i.e.,

$$[H^+][OH^-] = 10^{-14} \tag{4}$$

With this information it is now possible to construct an equilibrium line for phenol between the waste water at a given pH, say 7, and the emulsified caustic solution, say 10 wt% NaOH. Subscripts c and w in the following derivation will refer to concentrations in the caustic and waste water phases, respectively.

The derivation of the equilibrium line is based on the fact that at equilibrium the concentration of the undissociated phenol, which is the diffusing species, must be equal on both sides of the membrane.

For the caustic phase (10 wt% NaOH = 2.773 M), electrical neutrality requires that, disregarding the H^+ concentration,

$$[Na^+]_c = [OH^-]_c + [PhO^-]_c = 2.773 \qquad (5)$$

or

$$[OH^-]_c = 2.773 - [PhO^-]_c$$

which can be substituted into the water equilibrium (4) to give

$$[H^+]_c = \frac{10^{-14}}{2.773 - [PhO]_c} \qquad (6)$$

When this expression is used to replace the $[H^+]$ in the phenol dissociation equilibrium (3), one obtains the following expression for the undissociated phenol in the caustic phase:

$$[PhOH]_c = \frac{[PhO^-]_c \times 10^{-4}}{1.28(2.773 - [PhO^-]_c)} \qquad (7)$$

This equation shows that in the caustic phase phenol is essentially completely dissociated, and that for all intents and purposes the total phenol concentration equals the dissociated phenol concentration, or

$$[\text{total PhOH}]_c = [PhO^-]_c + [PhOH]_c \simeq [PhO^-]_c \qquad (8)$$

Turning now to the waste water phase, which is assumed to be at pH = 7, i.e., $[H^+]_w = 10^{-7}$, one obtains the following relationship for the phenate ion by again using the phenol dissociation Eq. (3):

$$[PhO^-]_w = \frac{1.28 \times 10^{-10}[PhOH]_w}{[H^+]_w} = 1.28 \times 10^{-3}[PhOH]_w \qquad (9)$$

In contrast to the caustic phase, practically all of the phenol in the waste water is therefore present in the undissociated form, or

$$[\text{total PhOH}]_w \simeq [PhOH]_w \qquad (10)$$

Now, at equilibrium, since undissociated phenol is the species diffusing through the oil membrane, it must be at the same concentration in the caustic and waste water phases, or

$$[PhOH]_w = [PhOH]_c \qquad (11)$$

It is therefore possible to substitute expressions (11), (7) and (8) into Eq. (10) to obtain the equation for the phenol equilibrium distribution curve:

$$[\text{total PhOH}]_w \simeq \frac{[\text{total PhOH}]_c \times 10^{-4}}{1.28(2.773 - [\text{total PhOH}]_c)} \tag{12}$$

Equation (12) is the equilibrium curve which relates the concentration of phenol in the caustic with that in the waste water at pH = 7. The relationship is plotted in Fig. 6, showing that phenol can exist in caustic in equilibrium with waste water even if the concentration of the phenol in the caustic is 10,000-fold that in the waste water.

While this concentrating effectiveness of liquid membrane emulsions is a very desirable and important feature in waste water treatment, the clean-up which can be achieved must next be considered. This can be done easily with the help of the equilibrium curve.

Assume the waste water starts out with 200 ppm phenol. In a single treat, if the caustic is spent to about 50%, no better than an exit phenol concentration of 7.4 ppm can be achieved at equilibrium. Allowing for a reasonable driving force will therefore leave 15 to 25 ppm phenol in the effluent water.

By adding a second and, if required, third contacting stage, the phenol can be cleaned up to a level between 0.25 and 2.5 ppm. Countercurrent flow of emulsion and feed water is necessary to achieve satisfactory chemicals utilization.

The design of a typical countercurrent clean-up system is illustrated in the stage-to-stage diagram of Fig. 7. The operating line of slope (lb waste

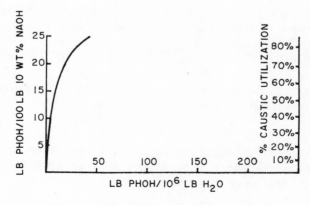

FIG. 6. Phenol equilibrium: caustic vs water.

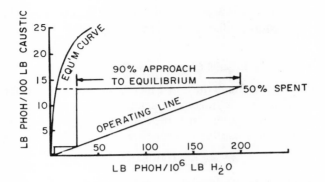

F$_{IG}$. 7. Stage-to-stage diagram.

water/lb of emulsified caustic solution) is the result of a material balance over the unit. In estimating stage requirements, a 90% approach to equilibrium and below 0.5 ppm phenol in the effluent were assumed. Also, a 50% spent caustic was taken as the desirable maximum utilization.

Similar equilibrium curves and operating lines can be constructed for other systems, i.e., removal of HCN, H_2S, NH_3, etc., from waste water into the appropriate emulsion. In general, the smaller the dissociation constant, the fewer stages will be needed to accomplish a given clean-up.

What is particularly interesting with regard to this removal technique as applied here is that it is so selective in its removal of only that or those materials which (a) can pass through the membrane, and (b) will react with the reagent inside the emulsion and form a nondiffusing product.

In the case of phenol, other weak acids which happen to be present will compete with phenol in permeating through the membrane and reacting with the encapsulated caustic. Thus acetic acid, H_2S, and HCN, all of which have small dissociation constants and good solubility in the organic membrane layer, will permeate along with the phenol and accumulate on the inside. However, strong mineral acids, on account of their high degree of dissociation and low solubility in the oil phase, will not tend to diffuse across the oil membrane. The net result is that Cl^-, SO_4^{2-}, etc. are effec-tively excluded from the encapsulated caustic. Thus the method can be used to remove phenol even in the presence of large amounts of chloride.

The presence of bases such as NH_3 and amines does not interfere with the removal of phenol. These materials are quite soluble in the oil film, will diffuse across, but since they do not react they will rapidly build up to about the same concentration inside and outside the liquid membrane, and all further permeation will come to a halt.

Rate Equations for Liquid Membrane Permeation

Permeation rates through the liquid membrane have to be determined to allow sizing of contacting stages and agitation requirements.

The rate of permeation of constituent A was defined earlier by the basic rate Eq. (2), repeated here for convenience:

$$dN/d\theta = (D)(\text{area})(\Delta c/\Delta x) \tag{2}$$

Since the area available for permeation and Δx are difficult to measure for a liquid membrane system, the group $(D)(\text{area})/\Delta x$ can be replaced by $D'(V_E/V_w)$, where D' is an effective permeation rate constant and V_E/V_w is the treat ratio, e.g., volume of emulsion/volume waste water. Essentially, Δx has been lumped into the permeation rate constant, and it has been assumed that when the emulsion breaks up into drops, the area available for permeation is proportional to the amount of emulsion used in the treat per unit of waste water.

Care has to be taken in the handling of the Δc term. Only the concentrations of the un-ionized species must be used in computing this concentration differential, since this is what propels the materials across the membrane, not the difference in total concentration. Thus, in the case of phenol removal from 200 ppm in the waste water into a 10 wt% caustic membrane emulsion, consider the point when the caustic solution is 10% spent, i.e., contains 2.4 lb phenol/100 lb caustic solution. While this represents 23,500 ppm (wt) total phenol content, the concentration of undissociated phenol in the caustic is only 1/30,000 of this value (see Eq. 7), or less than 1 ppm. On the other hand, since the phenol in the pH 7 waste water is essentially completely undissociated (see Eq. 10), all the 200 ppm can be taken as undissociated phenol concentration. Therefore, the driving force for phenol across the membrane is $200 - 1 = 199$ ppm or 0.00211 gmole/liter.

The flux of permeating species in the rate equation, $dN/d\theta$, while calculated as undissociated species, is, however, measured as total concentration (ionized plus un-ionized), since there is rapid interconversion between the dissociated and undissociated forms in the aqueous phases, depending on pH.

Permeation rates have been measured for various systems, both in the laboratory and in a pilot plant, and reasonably good correlation between

TABLE 2

LMP Rate Constants for NH_3

Time interval (min)	Oil/aqueous, D' (min^{-1})			
	2/1	1.2/1	0.8/1	0.6/1
0–1	0.60	0.20	0.75	0.99
1–4	0.36	0.48	0.75	0.71
4–9	0.51⎫		0.71	0.51
9–19	0.65⎭	0.71	0.85	0.99
Average	0.53	0.53	0.77	0.80

TABLE 3

Phenol Removal by LMP ($V_E/V_W = 266/500 = 0.532$)

Interval (min)	θ (min)	c_{in}/c_{out} (ppm/ppm)	D' (min^{-1})
0–2	2	1060/61	2.7
2–5	3	61/4.1	1.7
5–18	13	4.1/3	~0
0–5	5	1060/4.1	2.1

continuous pilot unit data and batch runs in a beaker have been achieved. Laboratory data were used to calculate the permeation rate constants for both ammonia and phenol listed in Tables 2 and 3, respectively.

The phenol rates are higher but indicate a marked drop-off as phenol removal approaches completion, while the D''s for ammonia seem to remain reasonably constant over the period of the test. The ammonia rates appear to increase with decreasing oil/aqueous ratio in the emulsion formulation. This is to be expected, since lower oil/aqueous implies thinner membranes, higher diffusion rates.

It should be noted that the oil/aqueous ratio of the emulsion is just one of the variables which affect permeation rate. Others are temperature, mixer design and speed, membrane formulation, vessel design, scale of operation, and many other factors.

The LMP rate constants in Table 2 were calculated from the laboratory data by using the batch permeation equation

$$\ln \frac{c_{in}}{c_{out}} = (D')(V_E/V_w)(\theta) \tag{13}$$

where c_{in} and c_{out} are the initial and final concentrations of the material being removed by LM permeation, respectively, $V_E' V_w$ is the treat ratio, θ is the contacting time for any given interval, and D' is the permeation rate constant.

Equation (13) is easily derived from the generalized permeation Eq. (2).

Once the LMP constant D' has been determined, continuous cocurrent plant-type treating equipment can be designed on the basis of Eq. (14), (15), (16), and (17) listed below. Equation (15) is for a single, well-stirred continually fed reactor for the case where there is no appreciable solute concentration in the emulsion.

For continuous systems, define

$$K = \frac{v_w}{D'}\left(\frac{v_w}{v_e} + 1\right)$$ (14)

where v_w is the water rate and v_e is the emulsion rate, both in liters/minute. Then the reactor volume V_R is given by

$$V_{R_1} = K\left(\frac{c_{in}}{c_{out}} - 1\right) \qquad \text{for a 1-stage system} \qquad (15)$$

$$V_{R_2} = 2K\left(\sqrt{\frac{c_{in}}{c_{out}}} - 1\right) \qquad \text{for a 2-stage system} \qquad (16)$$

$$V_{R_n} = nk\left(\sqrt[n]{\frac{c_{in}}{c_{out}}} - 1\right) \qquad \text{for a } n\text{-stage system} \qquad (17)$$

Since the permeation rate is controlled by the effluent concentration, which is low, a multistage cocurrent mixer results in a much smaller required mixer volume. It can easily be shown that for a 2-stage system the minimum vessel volume is obtained when both stages are of equal size, and this is reflected in Eq. (16) for the 2-stage mixer. For the n-stage cocurrent system, application of the 2-equal stage optimum to any two consecutive stages leads to the obvious conclusion that it is best to make all n stages of equal volume. This is the basis for Eq. (17).

Transfer of Ions

The present discussion covers the permeation of phenol, ammonia, H_2S, or other somewhat oil-soluble molecular species which are converted to a predominantly oil-insoluble ionic species by reaction with a base or acid inside the emulsion droplets. It has been possible to remove ionic contaminants from waste water by the addition of solubilizing agents, such as ion exchange compounds and complexing agents, to the oil phase of the membrane emulsion. Then, by maintaining a proper pH differential between the waste water and the emulsified water droplets, respectable permeation rates and concentration build-up of the contaminating metal ions into the emulsion could be achieved.

CONCLUSION

Design and economic studies have been made for a number of systems, and they were found to compare reasonably well with conventional clean-up methods. Particularly when the amount of material to be removed is present in low concentration, the chemicals consumption is correspondingly low and the liquid membrane technique could be a worthwhile alternate for investigation. It can be particularly useful where other methods fail due to the presence of interfering substances, very high or low pH's, or some other special circumstance. In any case, the removal of contaminants from waste waters by liquid membrane permeation is a novel technique which is being actively pursued at the present time.

REFERENCES

1. N. N. Li, U.S. Patent 3,410,794 (November 12, 1968).
2. N. N. Li, R. P. Cahn, and A. L. Shrier, U.S. Patent 3,617,546 (November 2, 1971).
3. N. N. Li and A. L. Shrier, *Recent Developments in Separation Science*, Vol. 1, Chemical Rubber Co., Cleveland, Ohio, 1972, p. 163.
4. N. N. Li and R. P. Cahn, U.S. Patent 3,719,590 (March 6, 1973).

The Plasma Chromatograph as a Separation–Identification Technique

MICHAEL M. METRO and ROY A. KELLER

DEPARTMENT OF CHEMISTRY
STATE UNIVERSITY OF NEW YORK
COLLEGE AT FREDONIA
FREDONIA, NEW YORK 14063

Abstract

The plasma chromatograph is evaluated as a separation and identification device using binary mixtures of ketones. Discrimination of components is not encouraging and the ion–molecule spectra of a mixture is not always the sum of the spectra of the individual components.

INTRODUCTION

Operating Principle

The Plasma Chromatograph, PC (Franklin GNO Corp., PO Box 3250, West Palm Beach, Florida, 33402) is an ion-molecule/ion-drift time of flight mass spectrometer. Its construction and operation has been described (1–4). In brief, a pure gas, e.g., nitrogen (Ultra Pure) or air (Zero Air) of low water content (2 to 10 ppm) (4), is passed over a ^{63}Ni source. The emitted β-electrons produce secondary electrons and counterions by inelastic collision and are soon slowed to thermal energies. The electrons are captured by electronegative species when present in the gas. Keller and Metro (4) reviewed the identity of these *primary reactant ions*. They are of the general type $(H_2O)_nH^+$ and $(H_2O)_nNO^+$ for positive ions and $(H_2O)_nO_2^-$ and $(H_2O)_n(CO_2)O_2^-$ for negative ions. Others have been suggested, e.g., CO_3^-. The degree of hydration, n, and the relative amounts of each species depends upon the composition of this gas, the *carrier* or *reactant gas*, particularly the water content, and the temperature of the chamber in which this occurs, the *reaction chamber*. A constant but adjustable voltage exists across this region and the following drift region. This field may be reversed so either positive or negative ions flow down the axis of the tube. Because the process occurs at atmospheric pressure, constant ion velocities are soon produced, i.e., viscous flow. The reaction chamber terminates at an electronic gate which is pulsed to admit the mixture of ion-molecules to the *drift region*. Gas, the *drift gas*, nitrogen, flows counter to the ion beam. The intent is to prevent uncharged sample molecules from entering the drift region and participating in further charge exchange with the ion-molecules. Thus only ions enter the drift region. Keller and Metro (4) argue that this is differential migration from a narrow zone, the gated pulse, and hence the process is a form of chromatography. Each pulse of species is sensed by an electrode at the terminus of the drift tube.

Three read-out systems are available. Another gate, which is controlled to open at an increasing and regulated delay time after the first gate, is positioned in advance of the sensing electrode. Thus each admitted pulse is scanned for a signal at a particular but different delay time. These signals are sent to an x-y recorder and a curve of current intensity vs delay time read out. Two minutes is a reasonable time to scan 20 msec. This we have abbreviated as PC-mg (moving gate). Because of the excellent GNO high speed electronics, a single pulse may be scanned with the second gate open and the signal observed on an oscilloscope, PC-os. We use a storage scope and photograph the scans. This signal may also be sent to a signal averaging computer, stored, and then displayed on an oscilloscope or an x-y recorder, PC-sac. In terms of response, in order of superiority, PC-sac > PC-mg > PC-os.

When a sample is introduced into the reaction chamber, it interacts with the reactant ion-molecules by charge transfer to produce a new set of *product* ion-molecules which are admitted to the drift region for separation.

The overwhelming attraction of the PC is its calculated and experimentally demonstrated response to 10^{-12} mole fraction of some materials (4). This paper will examine experience with the instrument as a separation device.

Response

The β-source produces a constant stream of ionizing electrons. Each of these produces about 10^3 secondary electrons and counterions before they reach thermal energies. The current established in the reaction chamber is about 1.5×10^{-9} A in a 1-cm diameter beam. If $_0n_R$ is the concentration of reactant ions in the reaction chamber then

$$_0n_R = n_R + n_P \tag{1}$$

where, after the sample is introduced, n_R is the concentration of remaining reactant ion-molecules and n_P is the concentration of sample or product ion-molecules. This is a charge conservation condition and sets an upper limit on the response. We suggest that an *appropriate sample* is one less than or just equal to a size where $n_R = 0$. Any sample above this we have termed an *overload*. Overload will occur at different sample sizes of different substances. The nature of the product ion-molecules and their relative distribution will depend upon the competitive Lewis acid/base interactions which are both temperature and concentration dependent. Cram and Chesler (5) found some Freons different by as much as 280 in response factors.

Sampling

Pulse sampling is the injection of a finite amount of material over a short period of time into the carrier stream. The sample encounters a large pristine adsorptive surface. Karasek (6) found that he could record plasmagrams of the polychlorinated biphenyls for *several hours* as the sample slowly escaped from this surface. Cram and Chesler (5) strongly advise silylation of the grids and glass surfaces to reduce the clearing time of the sample and improve the response.

Most experience has been with overload pulse sampling. In time, as the sample concentration drops, the reactants reappear at the expense of the product ion-molecules. Because of the charge conservation, to a first approximation, the total area under all peaks on the plasmagram should remain constant. This is not strictly true because the constant admitting gate width will admit a larger number of high velocity ion-molecules. The faster ions will account for a proportionally larger admitted current. Thus the total peak area for a population of predominantly fast ion-molecules should be greater than the total area for a population of slow ion-molecules. Peak areas will be both concentration and velocity dependent. Other time-of-flight mass spectrometers approach this problem by placing a grid ahead of the admitting gate which is pulsed just prior to opening the admitting gate. This grid mixes the various "velocity" species. The admitted pulse of ions are velocity separated only after admission to the drift region.

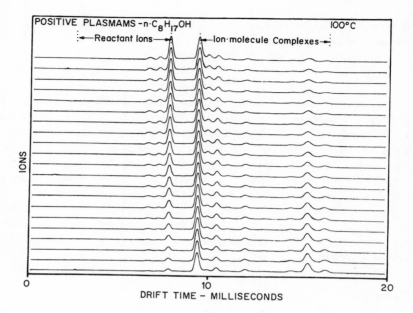

FIG. 1. Successive PC-mg scans of 1-octanol (7). (Courtesy of *Journal of Chromatographic Science.*)

It is advised that plasmagrams are reliable only after the reactants begin to reappear. Figure 1 is the plasmagram (PC-mg) of a 1-octanol pulse sampled at overload and recorded after reactants were apparent (7). It is a common and acceptable result. We note that as the reactants recover, the large drift time peak diminishes while the major peak remains fairly constant. It is important to note that peak position does not change. Figure 2 (8) is a PC-os scan of a severe overload pulse of di-*n*-butyl ether. The scan interval is the time (in seconds) between photographing the oscilloscope display. The instability is obvious. Results for diethyl and di-*n*-propyl ethers were similar. Reactant species reappeared after very different time periods for the three ethers, and the drift times for the product ions present at that instant showed no correlation with molecular weight. Speculating on the drifting "apparent" peak, Keller and Metro (4, 8) suggested that because of the response of the PC, all samples must be considered as mixtures, and that primary component and impurities constantly change in concentration in the reaction chamber as they adsorb and desorb on the surface at different rates to give species of fixed drift

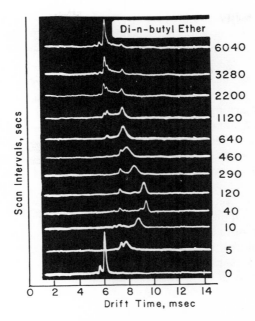

FIG. 2. PC-os scans of di-*n*-butyl ether (*8*). (Courtesy of *Journal of Chromatographic Science*.)

times but constantly changing intensities. The drifting peak could be an envelope of these poorly resolved species. There is also strong evidence (*4*) that even for reactant species the ion-molecule pulse does not contain a single species. Cram and Chesler (*5*) suggested that although neutral sample molecules are excluded from the drift region, ion-molecule reactions with materials in the drift gas could lead to continued reaction in this drift space. This could lead to further reaction, e.g., the following reactions suggested by Griffin et al. (*9*):

$$MH^+ + H_2O + N_2 \rightleftarrows M(H_2O)H^+ + N_2 \qquad (2)$$

$$M_2H^+ + N_2 \rightarrow MH^+ + M + N_2 \qquad (3)$$

when carrier and drift gas are identical. Metro and Keller (*8*) recalled the earlier work of Keller and Giddings (*10*) who showed that for A ⇄ B, where the elution times of A and B are presumed different, then if reaction rates are sufficiently fast that each molecule can undergo several transitions in the period of separation, A and B *cannot be separated* and a single zone will elute with an elution time between those proposed for A and B. Since reaction rates are concentration dependent, overload would enhance this effect. The drifting "apparent" peak could also be due to such continued reaction.

The most clear-cut case of appropriate pulse sampling is the recent work of Cram and co-workers (*11*) with an exponential dilution flask. A heated 1 liter flask was arranged so that the carrier gas could sweep through it continuously at a temperature sufficiently high that the sample was in vapor form. The outlet was connected to a 3-way valve which could vent the sample or direct it into the PC. A quick turn of the valve sent a pulse sample to the PC. By waiting an appropriate period of time, the concentration of vapor could be reduced to the level of an appropriate sample. They used two of Metro and Keller's ether samples with a PC-sac. With overload pulse sampling they observed much the same thing as Metro and Keller; with appropriate pulse sampling they achieved stable ether spectra. Appropriate samples may simplify the plasmagrams by avoiding high molecular weight ion clusters which appear at high sample concentrations. Hopefully, a report of their work will appear soon.

An appropriate pulse sample seems superior to an overload pulse. The question of interacting species still remains but may be of reduced importance because of the low sample concentrations. There are some apprehensions about its use with mixtures. (a) The sampling flask must be at a sufficiently high temperature and adsorptive effects sufficiently moderated that the vapor composition is identical to that of the original sample and remains that way from the time of injection of the sample into the flask and the time of admitting a pulse to the PC. Because the vapor concentration is constantly reduced in the flask, it is doubtful if a reliable equilibrium period exists. (b) Although vapor/adsorbate equilibrium may exist in the dilution flask, the sample will still encounter a pristine surface in the interface between the flask and the PC injection port, injection tube, reaction chamber, and drift tube. A new equilibrium will be established which may affect the vapor composition. We fear that a highly adsorptive minor constituent can be so reduced in such a small sample that it is not seen by the PC. Elevated temperature coupled with surface deactivation, e.g., silylation, seems mandatory. (c) Cram (*11*) estimates sample sizes to be in the picog level. If the sample contains a relatively unresponsive minor constituent, it may remain undetected.

We have elected to do pulse sampling with overload with binary mixtures, where concentrations are not minor, to identify any promise as a separation-identification technique.

Steady-state or *continuous sampling* is introduction of the sample over a sufficient period of time that equilibrium is established between adsorbate and vapor throughout the PC so the vapor composition is constant and is that of the sample. Steady-state sampling with overload has not been explored. Appropriate steady-state sampling has been approximated. Karasek and colleagues (*12, 13*) drove alcohol vapors from a syringe slowly into the PC but seem to have terminated injection when the reactant ions vanished. Horning et al. (*14*) and Griffin et al. (*9*), using relatively nonvolatile samples, raised the temperature of the sample exposed to the carrier stream until the vapor concentration was such that reactant ions were on the point of vanishing. Unfortunately, they did not use the total PC system; only the reaction chamber as an ionization source to a vacuum mass spectrometer. Such sampling is unsuited for mixtures since the vapor composition is dependent upon the vapor pressure of the pure components and their concentration in the liquid mixture. Coffey (*15*) has perhaps come the closest. He introduced ammonia vapor at about 0.03 ppm and allowed his sample system to reach a steady state for at least 24 hr before taking data. Because of the high proton affinity of ammonia, species were exclusively $NH_4^+(H_2O)_n$.

Ideally, one requires a completely vaporized sample of very small amount and of unchanging concentration but of large total volume which can be steadily introduced over a relatively long period of time.

Product Ion Species

Unlike other chromatographic separations, a single molecular species will most likely yield more than one product ion-molecule. This is very undesirable to the separations chemist. If M is the sample species, it is possible to produce (*4*): (a) ion clusters $(M)_m H^+$, $(M)_m^-$, $(M)_m O^-$, and $(M)_m O_2^-$; (b) hydrated ion clusters $(M)_m (H_2O)_n H^+$ and $(M)_m (H_2O)_n O_2^-$; and (c) fragments, e.g., halides X^- from some halogenated aliphatics and aromatics and rupture at ether linkages with the Freons. Higher molecular weight clusters are apparent with large samples. The nature and relative distribution of the species are much dependent upon sample concentration and temperature. Add to this the strong possibility of continued reaction in the drift region, and the situation does not seem propitious. It is concievable that mixtures (components M and N) could give mixed species $M_m N_n H^+$ and hydrates, and that fragments could recombine. Karasek and Kane (*13*) found: EtOH, 2 peaks; 1-BuOH, 5 peaks; 1-HexOH, 4 peaks; and 1-OctOH, 6 peaks. They remark that for identification purposes the temperature must be carefully selected, some indication of concentration should be available, and the samples should be a single component.

Mixtures

Unfortunately, there has been very little experience with deliberate mixtures, and mixtures are the subject of separations. Karasek (*16*) showed a relatively uncomplicated plasmagram of dimethylsulfoxide, malathion, and triethylphosphite. Carroll (*17*) reported good resolution of a 3-component mixture of chlorodibenzodioxins. Karasek and Keller (*18*) found the plasmagram of musk ambrette to be simple and independent of the solvent and of column bleed when a PC was interfaced with a gas chromatograph. Horning et al. (*14*) introduced samples as solutions, i.e., binary mixtures. The solvent, benzene or chloroform, present in excess, behaved as if it reacted directly with the source β-electrons to form a new set of reactant species to introduce a different series of Lewis acid/base interactions with the sample species. Cram and Chesler (*5*) investigated a binary mixture of Freons and concluded that a plasmagram of a binary system is very difficult to interpret by itself.

Identification

The hope that the masses of product ion-molecules could be determined from a general reduced mobility vs mass calibration curve, first established by interfacing a PC with a mass spectrometer, has proved chimerical. The very careful work of Griffin et al. (*9*) established that, for unknown compounds, the standard error in determining mass by such a curve is 20%. The situation is apparently better for members of a homologous series (2%). Masses can only be reliably determined by interfacing the PC with a mass spectrometer. It, too, is not problem-free, e.g., adiabatic expansion into a vacuum may alter species.

What remains seems to be a hope that each solute may have a relatively simple and unique plasmagram ["fingerprint" (*1*)] and that the plasmagram of a mixture is an additive sum of the plasmagrams of the components. This paper investigates that possibility using 2-butanone and 2-octanone both alone and together in deliberate mixtures to two different temperatures.

EXPERIMENTAL

Samples

The ketones were from the gas chromatographic reference compound kits prepared by Poly-Science Corp. (6366 Gross Point Road, Niles,

TABLE 1

Sample Composition

	2-Butanone		2-Octanone	
	Wt-%	Mole-%	Wt-%	Mole-%
Low temperature				
Mix 1	57.91	70.98	42.09	29.02
Mix 2	23.97	35.92	76.03	64.08
High temperature				
Mix 3	56.24	69.56	43.76	30.44
Mix 4	39.22	53.88	60.78	46.11

Illinois, 60648). Minimum purity of these references is 99.5%. Mixtures were prepared by weight in a septum-sealed sampling bottle. These were prepared and used immediately to minimize changes in composition on standing. Table 1 shows the composition.

Sampling

Samples were not introduced into the PC until the reactant ion spectrum was that associated with a clean tube. Thus at least 24 hr elapsed between experiments.

A Hamilton 1 μl syringe (7001 series) was flushed with the liquid sample 5 to 7 times, the last liquid expelled, the plunger drawn back to 0.5 μl, and the material remaining injected into the PC. This was still an overload. Presuming that the greater portion of the sample was vapor, it is highly probable that the vapor composition did not represent the solution composition. The syringe was cleaned between injections by drawing air through the heated needle for 5 min.

Data Collection

A PC-os and PC-mg scan was taken of the reactant ions. A PC-os scan was taken immediately after sample injection and this continued until the spectra showed promise of the reactant species. Thereafter PC-os and PC-mg scans were taken. We note from the PC-os scans that there was no drifting apparent peak as observed with severe overload pulse sampling of the ethers (8). Peak intensities changed but not drift times. Following the persistent advice that spectra are only meaningful when reactant species begin to reappear, we considered only the PC-mg scans where reactants are apparent and show the PC-os scans immediately preceding and immediately following this PC-mg scan. The time interval from the instant of injection is shown in the figures. We report reduced mobilities *(4)* of persistent peaks. By this we mean: If a peak appears on several PC-os scans before and after the appearance of the reactant species, we report its reduced mobility even though it may appear as a minor peak on the PC-mg scan selected.

PC Conditions

Table 2 shows the PC conditions.

As with the ethers *(8)*, the PC housing temperature was sensed with a Dymec Model 2081A quartz thermometer probe, insulated with glass wool, placed flush against the metal housing at the flange and bolt face.

RESULTS AND DISCUSSION

Figures 3 through 10 show the PC-mg scan where reactants just appear and the immediately preceding and following PC-os scan. The time the scans were made after sample injection are shown on the scans. The PC-mg scan times were 1.75 min. The peaks are labeled with raw drift times in milliseconds, i.e., not corrected for temperature. The advantage of the immediately following PC-os scan is that comparison of it with the PC-mg

TABLE 2

Operating Parameters for the Plasma Chromatograph

Flow rates:		
	Drift gas:	Ultra pure grade nitrogen, 500 ml/min
	Carrier gas:	Ultra pure grade nitrogen, 100 ml/min
Temperature:		
	Carrier inlet:	138 °C
	PC housing:	107–109 °C (low temperature)
		152 °C (high temperature)
	Drift gas inlet:	129 °C
Voltage:		
	3500 V, positive mode (collection of positive ions)	
Gates:		
	PC-mg:	Admitting gate, 0.2 msec; exit gate, 0.2 msec
	PC-os:	Admitting gate, 0.2 msec; exit gate, open

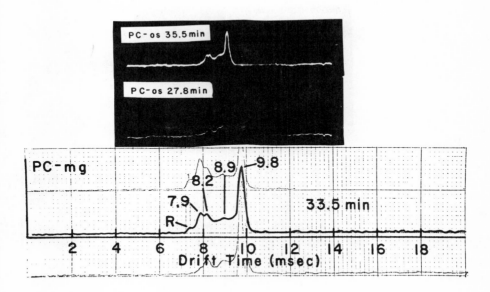

FIG. 3. PC-mg and PC-os scans of 2-butanone at low temperature (107.9 °C).

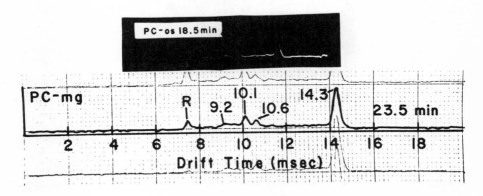

FIG. 4. PC-mg and PC-os scans of 2-octanone at low temperature (106.6 °C).
The PC-os scan following PC-mg is missing because of film failure.

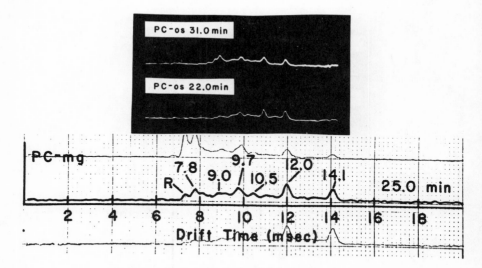

FIG. 5. PC-mg and PC-os scans of Mixture 1 (58/42 B/O) at low temperature (109.3 °C).

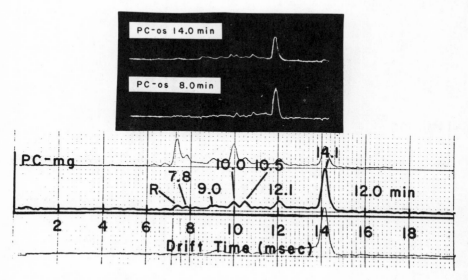

FIG. 6. PC-mg and PC-os scans of Mixture 2 (24/76 B/O) at low temperature (108.0 °C).

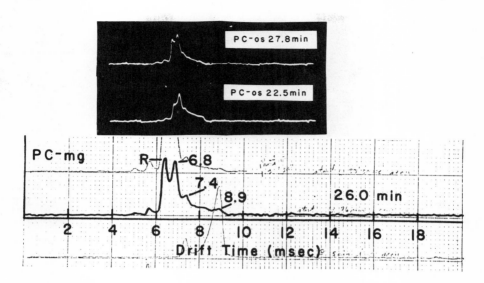

FIG. 7. PC-mg and PC-os scans of 2-butanone at high temperature (152.4 °C).

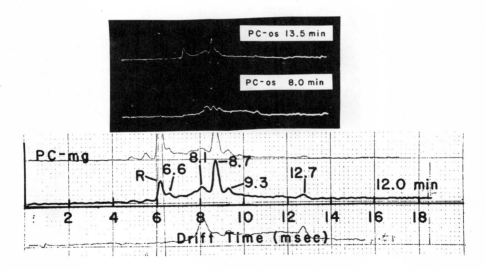

FIG. 8. PC-mg and PC-os scans of 2-octanone at high temperature (152.5 °C).

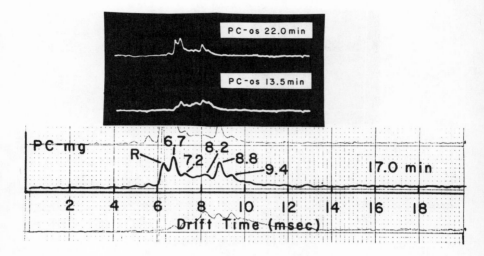

FIG. 9. PC-mg and PC-os scans of Mixture 3 (56/44 B/O) at high temperature (151.8 °C).

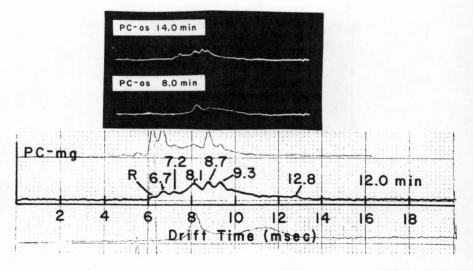

FIG. 10. PC-mg and PC-os scans of Mixture 4 (39/61 B/O) at high temperature (151.9 °C).

scan detects any dramatic changes which might have taken place during the PC-mg scan time. Unlike the ethers, no such evolution occurred. In all cases the PC-mg scan is superior to the PC-os scan, which is to be expected.

Each calibration on the PC-mg scan corresponds to 0.2 msec. Because of some electronic noise, temperature difference, and slight peak shifts likely arising from the peak representing an envelope of poorly resolved species, we suggest that peaks not differing by more than 0.2 msec are due to the same component or are not definitive of the sample. Table 3 shows the raw drift times of the pure ketones and their mixtures at lower temperature. The table is arranged vertically in order of increasing drift time, and peaks of drift times within 0.2 msec are on the same horizontal line. Table 4 represents the results at the higher temperature.

TABLE 3

Low Temperature Raw Drift Times (msec)

2-Butanone, 107.9 °C	2-Octanone, 106.6 °C	Mix 1, 58/42 B/O, 109.3 °C	Mix 2, 24/76 B/O, 108.0 °C
7.9		7.8	7.8
8.2			
8.9		9.0	9.0
	9.2		
9.8		9.7	
	10.1		10.0
	10.6	10.5	10.5
		12.0	12.1
	14.3	14.1	14.1

TABLE 4

High Temperature Raw Drift Times (msec)

2-Butanone, 152.4 °C	2-Octanone, 152.5 °C	Mix 3, 56/44 B/O, 151.8 °C	Mix 4, 39/61 B/O, 151.9 °C
6.8	6.6	6.7	6.7
7.4		7.2	7.2
	8.1	8.2	8.1
8.9	8.7	8.8	8.7
	9.3	9.4	9.3
	12.7		12.8

We note in Table 3 that the 7.9 and 8.9 msec peaks for 2-butanone are definitive, i.e., they appear in both mixtures. The 9.8 msec peak is not; it appears only in the mixture where the butanone concentration is high. Likewise the 10.6 and 14.3 msec 2-octanone peaks are definitive since they appear in both mixtures. The 9.2 msec peak does not appear in either mixture, and the 10.1 msec peak appears only in the mixture of higher octanone content. Of particular interest is the 12.0 msec peak (Fig. 5 and 6) which appears in the mixture but never in either pure ketone. A careful examination of the PC-os and PC-mg scans of the pure ketones taken from the instant of injection up to and far beyond the reappearance of the reactants *show no evidence of a 12.0 msec peak for either ketone.* This peak is characteristic of the mixture and may be a mixed species as suggested earlier.

Table 4 shows that the only definitive peak for 2-butanone is the 7.4 msec peak. The 8.1 and 9.3 msec peaks are definitive for 2-octanone. The 12.7 msec peak appears only at the higher octanone concentration. There are no peaks present in the mixture which do not appear in the individual ketones. The 6.8 and 8.9 msec peaks are useless.

Our first conclusion is that "fingerprinting" is better at the lower temperature. We also conclude that the spectra of mixtures are largely the additive sum of the spectra of the components. The exception is the low temperature peak for the mixture which does not appear in the spectra of the pure components.

In order to compare results at different temperatures, we have calculated reduced mobilities, i.e., the speed of an ion in an electric field of 1 V/cm in a gas at standard conditions, by (4)

$$K_0 = (1/t)(D^2/V)(p/760)(273.2/T) \tag{4}$$

where t = observed transit time in seconds at p, T, and V
D = drift space distance, 5.96 cm
V = potential across the drift tube, 3500 V
p = pressure, 760 Torrs
T = drift space temperature, °K

Our results are shown in Table 5. The raw drift times are shown paranthetically to make it easier to compare Table 5 with Tables 3 and 4. We conclude:

1. The 0.96–0.99 cm/sec-V peaks are not useful. They appear for both ketones and two of the mixtures only at the higher temperature.
2. We feel that the 0.91–0.93 cm/sec-V peak for 2-butanone is definitive

TABLE 5

Reduced Mobilities ($\times 10$)

2-Butanone		2-Octanone		Mix 1, 58/42 B/O, 109.3°C	Mix 2, 24/76 B/O, 108.0°C	Mix 3, 56/44 B/O, 151.8°C	Mix 4, 39/61 B/O, 151.9°C
107.9°C	152.4°C	106.6°C	152.5°C				
9.2 (7.9)	9.6 (6.8)		9.9 (6.6)	9.3 (7.8)	9.3 (7.8)	9.7 (6.7)	9.7 (6.7)
8.9 (8.2)	8.8 (7.4)					9.1 (7.2)	9.1 (7.2)
8.2 (8.9)		7.9 (9.2)	8.0 (8.1)	8.1 (9.0)	8.1 (9.0)	8.0 (8.2)	8.1 (8.1)
7.4 (9.8)	7.3 (8.9)		7.5 (8.7)	7.5 (9.7)	7.3 (10.0)	7.4 (8.8)	7.5 (8.7)
		7.2 (10.1)	7.0 (9.3)	6.9 (10.5)	6.9 (10.5)	6.9 (9.4)	7.0 (9.3)
		6.9 (10.6)		6.0 (12.0)	6.0 (12.1)		
		5.1 (14.3)	5.1 (12.7)	5.1 (14.1)	5.2 (14.1)		5.1 (12.8)

at both temperatures. We admit that it does not appear for the pure ketone (Fig. 7), but an examination of this figure demonstrates poor resolution at a raw drift time of 7.2 msec where the peak ought to appear.

3. The 0.88–0.89 cm/sec-V peak for 2-butanone is useless.
4. The 0.81–0.82 cm/sec-V for 2-butanone is promising. We understand why it is undetected for the pure ketone. An examination of Fig. 9 also indicates poor resolution. We note that there is a peak at 8.2 msec, but we have assigned it to 2-octanone in the next row of 0.79–0.80 cm/sec-V. We must admit that this next assignment is based more on wishful thinking and that the best conclusion is that the 0.81–0.82 cm/sec-V for 2-butanone is not sufficiently distinct from the 0.79–0.80 cm/sec-V for 2-octanone to be distinctive for either.
5. The 0.73–0.75 cm/sec-V peak is useless. It appears in both ketones and in only two mixtures. One may, in fact, conclude that the next row of 0.72–0.74 cm/sec-V ought to be combined with this one and both judged as not distinctive.
6. The 0.69–0.70 cm/sec-V peak is clearly distinctive for 2-octanone at both temperatures for all mixtures.
7. The 0.60 cm/sec-V is the proposed cross-product.
8. The 0.51–0.52 cm/sec-V peak for 2-octanone is nearly distinctive. It is absent from the 2-butanone-rich mixture at the higher temperature. This may be a sampling problem.

At best there is potentially one reduced mobility which is useful for the identification of 2-butanone and two reduced mobilities distinctive for 2-octanone. We must conclude that identification by fingerprinting binary mixtures is not very promising at our experimental conditions. Refinement of sampling may very well show improvement and needs to be investigated, but it is not obvious to us that this would be the case.

Acknowledgments

The authors are grateful to Dr. Charles Lincoln of the Department of Physics for the loan of his quality oscilloscope and to the National Science Foundation, Grant GP-31824, for financial support of this investigation.

REFERENCES

1. M. J. Cohen and F. W. Karasek, *J. Chromatogr. Sci.*, **8**, 330 (1970).
2. F. W. Karasek, W. D. Kilpatrick, and M. J. Cohen, *Anal. Chem.*, **43**, 1441 (1971).

3. F. W. Karasek, O. S. Tatone, and D. M. Kane, *Ibid.*, *45*, 1210 (1973).
4. R. A. Keller and M. M. Metro, *Separation and Purification Methods*, *3*(1), 207 (1974).
5. S. P. Cram and S. N. Chesler, *J. Chromatogr. Sci.*, *11*, 391 (1973).
6. F. W. Karasek, *Anal. Chem.*, *43*, 1982 (1971).
7. F. W. Karasek and D. M. Kane, *J. Chromatogr. Sci.*, *10*, 673 (1972).
8. M. M. Metro and R. A. Keller, *Ibid.*, *11*, 520 (1973).
9. G. W. Griffin, I. Dzidic, D. I. Carroll, R. N. Stillwell, and E. C. Horning, *Anal. Chem.*, *45*, 1204 (1973).
10. R. A. Keller and J. C. Giddings, *J. Chromatogr.*, *3*, 205 (1960).
11. S. P. Cram, Personal Communication.
12. F. W. Karasek, M. J. Cohen, and D. I. Carroll, *J. Chromatogr. Sci.*, *9*, 390 (1971).
13. F. W. Karasek and D. M. Kane, *Ibid.*, *10*, 673 (1972).
14. E. C. Horning, M. G. Horning, D. I. Carroll, I. Dzidic, and R. N. Stillwell, *Anal. Chem.*, *45*, 936 (1973).
15. P. E. Coffey, "Ion Molecule Reactions of Atmospheric Importance. Explosive Growth Reactions Induced by the $NH_4^+(H_2O)_n$ Cluster," Publ. No. 204, Atmospheric Science Research Center, State University of New York at Albany, Albany, New York, 1972.
16. F. W. Karasek, *Res. Develop.*, *21*(3), 34 (March 1970).
17. D. I. Carroll, "Plasma Chromatography of Chlorinated Bibenzo-*p*-dioxins," Technical Report F-11(a), Franklin GNO Corp., West Palm Beach, Florida, 1971.
18. F. W. Karasek and R. A. Keller, *J. Chromatogr. Sci.*, *10*, 626 (1972).

Clathrates in Analytical Separations

E. C. MAKIN

RESEARCH DEPARTMENT
MONSANTO POLYMERS AND PETROCHEMICALS
ST. LOUIS, MISSOURI 63166

Abstract

Recent advances in clathrate chemistry and their use for analytical and pre-parative chemistry are discussed. The effect of molecular geometry and substituent groups of guest species on the capabilities of various clathrating compounds is reviewed.

Due in part to improvement in methods available for structural analysis, particularly in x-ray crystallography, the science of clathrate chemistry has made rapid advances in recent years.

The use of clathrates in separation processes has received considerable attention but in relatively restricted application. Further, their potential application in the separation field, particularly in analytical and preparative techniques, has been barely scratched.

This paper will briefly cover the background of clathrate chemistry, the state of the art with respect to separation-analytical techniques, and a projection of the future of clathrates in these fields.

While there are numerous types of guest–host combinations including channels, cages, and layers, emphasis will be placed on layers and cages highly specific in their selective capabilities based on molecular size and shape and capable of being formed and dissociated under fairly mild operating conditions of temperature and pressure or relatively simple chemical change.

A classic example of molecular size selectivity is the inorganic zeolites or "molecular sieves." They function as sieves with the cage or hole size amenable to change on a specific silica–alumina matrix only by exchange-ing all or part of the cation associated with the pore structure.

In the clathrate case, some hosts adjust somewhat to the size and shape of the guest component. This may reduce selectivity in some instances but can provide a wide range of compounds and conditions for tailoring a clathrate to achieve a specific separation.

Further, clathrates by virtue of their capabilities of forming and dis-sociating under relatively modest changes in temperature and pressure offer a broad spectrum of analytical capabilities. At the same time they can provide extremely stable cage-like structures for retention and storage if desired. The hydroquinone complexes of certain rare gases are a case in point and are discussed in detail later.

Molecular compounds such as the gas hydrates, the hydroquinone complexes, and carbohydrate inclusion compounds have been known for decades. However, a basis for extrapolating the behavior and capabilities of a host–guest species has only recently become better understood, and then in a relatively restricted area.

For example, the benzene–nickel cyanide amine complex $[Ni[CN]_2 \cdot NH_3 \cdot C_6H_6]$ has been known since 1897 (1). Only in recent years has this chemistry been expanded via the use of other amines in place of ammonia to alter the geometry of the basic lattice structure. This chemistry is also discussed in some detail later.

On the other hand, the chemistry of urea and thiourea adducts was studied in great detail following the discovery of this reaction by Bergen in 1940 (2). The chemistry and application of the urea and thiourea adducts for separating alkanes, naphthenes, and their derivatives has been discus-sed in such detail in numerous publications and patents during the past 25 years that little repetition is needed here.

It should be noted, however, that the investigative activity was probably catalyzed, at least in part, by the interest of the petroleum industry in the potential practical applications of the chemistry to separating valuable hydrocarbon species from crude mixtures.

Urea would probably have been the separating tool of choice for normal paraffin recovery had it not been for the simultaneous development of zeolite molecular sieves for the same purpose. Ingenious engineering concepts in the use of zeolites also aided the tool of choice.

Recent advances in clathrate chemistry indicate that the possibilities with respect to the discovery of new structural types and the application of these compounds and those already known has only been scratched, and we are reviewing their potential in a manner not unlike the iceberg phenomena.

While considerable attention has been directed to the chemistry of these compounds, the methods of exploitation have not been studied as thoroughly.

This is particularly true in the field of liquid phase preparative chromatography as well as vapor phase chromatographic techniques. The possibility of applying a multichemical, multistage technique for class and isomer separation has not received the attention it deserves. We will try to rectify that shortcoming in this brief presentation.

There are available today a spectrum of clathrating agents for separating a wide variety of chemical species as well as geometric isomers. Clathrates may also serve as templates in polymerization or other reaction processes.

A brief description of several clathrating structures and their known or potential application follows.

CLATHRATE TYPES

The molecular geometry of clathrating compounds varies widely as illustrated by the structures indicated in Figs. 1–8.

In the hydroquinone case (Fig. 1), the sulfur dioxide compound was discovered in the preparation of hydroquinone by reduction of an aqueous solution of benzoquinone with sulfur dioxide (3).

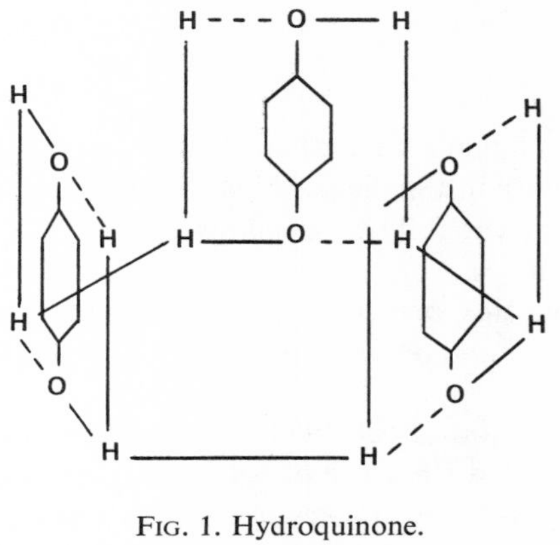

FIG. 1. Hydroquinone.

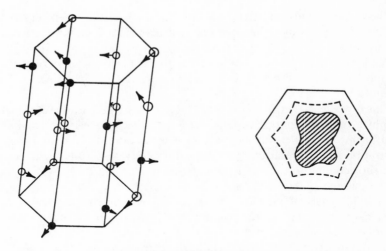

Fig. 2. Basic crystal lattice of urea complexes.

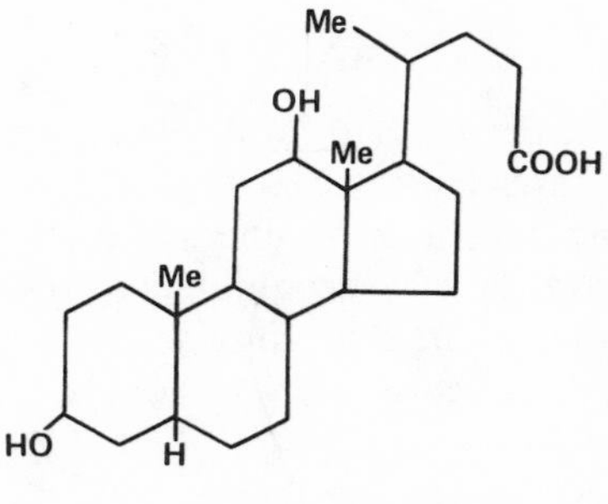

Fig. 3. Deoxycholic acid.

The urea case has already been alluded to in Bergen's discovery, as has the subsequent activity in this field which is continuing (Fig. 2).

Deoxycholic acid's capabilities in clathrating with *n*-alkanes has been known for years (Fig. 3).

An interesting question one could pose would be directed to the difference in this compound and urea, for example, that exhibit similar clathrating capabilities.

A more recent addition to this field is tri-*o*-thymotide, shown in Fig. 4.

In Fig. 5 the unrelated structure of 4,4'-dihydroxy triphenyl methane is illustrated. This is merely an illustration of clathrating compound types.

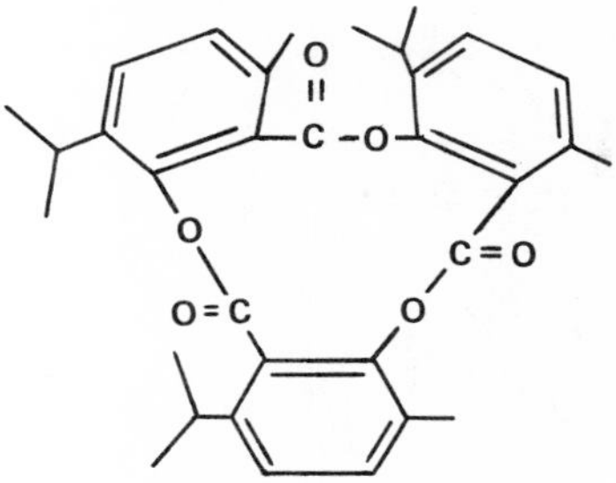

FIG. 4. Tri-*o*-thymotide.

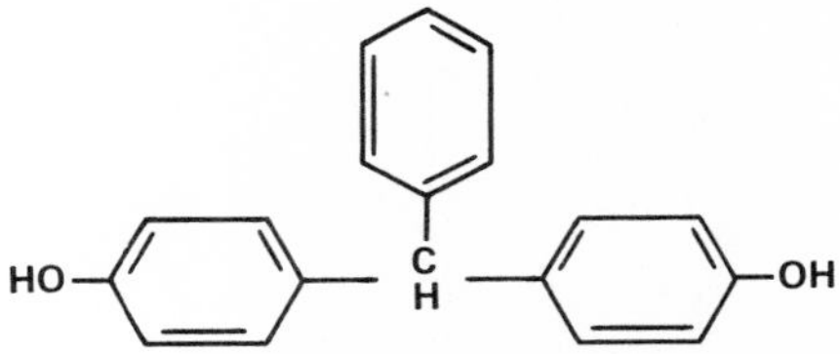

FIG. 5. 4,4'-Dihydroxy triphenyl methane.

One could draw an analogy with the hydroquinone complexes, but the extrapolation of this chemistry needs much more attention.

In Fig. 6 the well-known $Ni[CN]_2 \cdot NH_3$ complex is illustrated, based on the crystallographic studies of Powell and Rayner (4).

This chemistry has been expanded by a number of workers, particularly that of Schaeffer and co-workers (5), de Radzitsky and Hanotier (6), and Williams (7). Their work has concentrated essentially on a variety of Werner complexes comprised of the basic components depicted in Table 1.

These Werner complexes are also discussed in some detail later.

A more recent entry in the field is the compound shown in Fig. 7. Apparently a layered packing of this chemical forms channels and cages capable of clathrating with a wide variety of structures (8).

Cyclodextrins, polysaccharides, amylose, glycogen, amylopectin, and spirochromans (shown in Fig. 8) have clathrating capabilities of varying dimensions in cage and channel structures.

Briefly, this gives us an overview of the type of structures available for clathrate separations.

Many important structures such as hydrates, methyl naphthalenes,

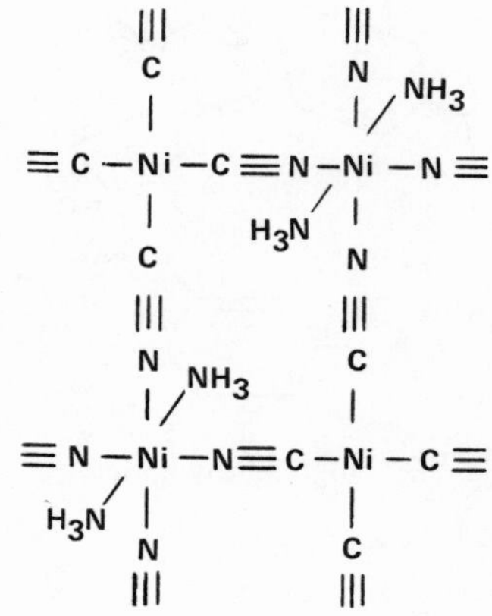

FIG. 6. Sheetlike structure of $Ni(CN)_2NH_3$.

TABLE 1

Werner Complexes

$M(N)_4X_2$

M = metal, Ni, Co, Fe, Mn, etc.
N = pyridene, substituted pyridines, benzylamines, etc.
X = thiocyanate, formate, halide (preferably SCN)

Selection of N compound determines the spacing
   of the metal complex layers

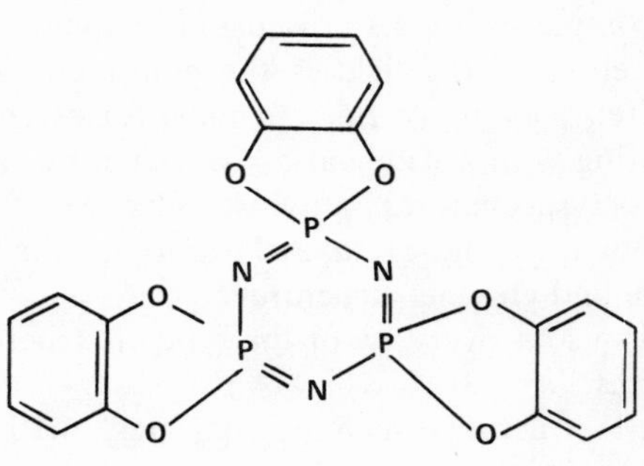

FIG. 7. Tris(o-phenylenedioxy)phosphonitrile trimer.

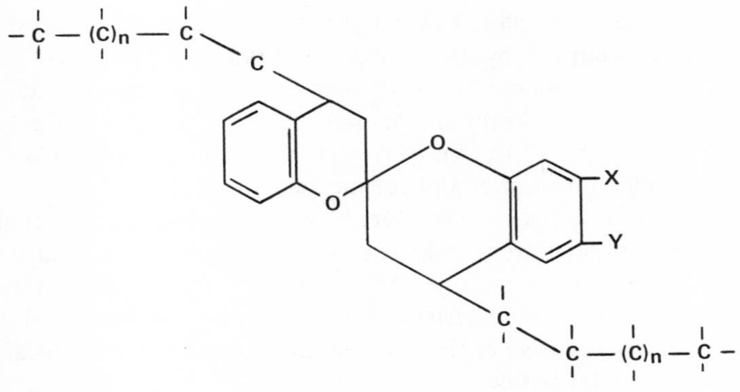

FIG. 8. Spirochromans.

substituted diphenyls, and other carbohydrates are also available for potential exploitation.

CLATHRATE GEOMETRY

Clathrates considered here exist in several geometric types including (a) layer structures, (b) channel or tunnel structures, and (c) cage structures.

As previously noted, the channel or tunnel type has received the greatest attention in recent years via in-depth investigations of the urea and thiourea complexes and their capabilities. This channel-type clathrate of urea involves a helix of molecules with six urea molecules per unit cell with hexagonal spacing. (Carbohydrates and proteins also form clathrates by wrapping around the guest compound.)

Hydrogen bonds tend to make the urea honeycomb cells contract: this force is resisted while the endocytic compound is present, but when the guest compound is liberated from the channel the hexagonal clathrate collapses to the less open and more stable tetragonal urea form.

In the layered structure case the metal nitrile–amine complexes are classics. The variation of the $Ni[CN]_2 \cdot NH_3 \cdot C_6H_6$ case is numerous. The distance between the metallic salt layers can be controlled by the choice of nitrogen containing complexing agent which, in turn, determines the distance between the floor and ceiling of the cage structure. To a degree, the wall dimensions have some latitude via the choice of substituent group on the chosen metal specie.

Cage structures are numerous, including the gas hydrates as well as the hydroquinone complex. In the hydrate case the voids in the structure are partially occupied by the parent clathrate former with additional pockets or cages available for occupancy by the guest compound.

Hydroquinone forms very stable clathrates with a variety of gases such as HCl, H_2S, SO_2, CO_2, Ar, Kr, and Xe, and low molecular weight aliphatics such as methanol and formic acid.

Clathrates of the gases can be stored for long periods of time at ambient temperature and pressure with little or no apparent vapor pressure exerted due to the tight cage structure. Pulverizing the crystals in a mortar and pestle develops the characteristic odor of certain encapsulated species, indicating partial release of the trapped component by mechanical breakdown of the crystal lattice.

These various geometric forms frequently overlap in their molecular sorting and encapsulating capabilities.

Table 2 compares the cage diameters of several clathrating species. The term "cage" is used here to denote the encapsulating geometry. As just pointed out, urea forms channel-type structures, while hydroquinone and the hydrate form cages with tri-*o*-thymotide exhibiting a series of cages connected by channels.

By the same token, the clathrate formers depicted in Table 3 also illustrate more than a single behavioral mode.

Attention is directed to the fact that the geometry of the cage formed is critical with respect to clathrating capabilities. Thus a channel structure such as thiourea can separate very high molecular weight isoalkanes, alkanes, and naphthenes whereas a cage structure such as the hydrates has a much more restricted functional molecular weight range of guest component.

TABLE 2

Cage Diameters of Several Clathrating Species

Host molecule	Cage diameter (Å)	Guest type
Hydroquinone	4.2–5.2	Ar: SO_2
Urea	5.25	*n*-Alkanes and derivatives
Thiourea	6.1	Isoalkanes, naphthenes *n*-alkanes > C_{16}
Desoxycholic acid	5–6	*n* + Isoalkanes, naphthenes
Tri-*o*-thymotide	4.8–6.9 (channel-cage)	*n* + Isoalkanes, naphthenes
Hydrates Type I	5.2–5.9	C_1–C_4 Paraffins and olefins
Hydrates Type II	4.8–6.9	

TABLE 3

Cage Diameters of Several Clathrating Species

Host molecule	Cage diameter (Å)	Guest type
4,4'-Dihydroxy triphenyl methane	6.0–6.5	n + Isoalkanes
$Ni(CN)_2 \cdot NH_3$	7.2	Benzene, thiophene, furan, pyrrole, aniline, phenol
$M \cdot (-N)_4 X_2$	Est. 7–10	Aromatic isomers
γ-Cyclodextrin[a]	9–10	Aromatics

[a] α and β Cyclodextrins accommodate n-alkanes and fatty acids.

Thus urea is highly specific due to its rigidity (inflexibility) and mode of formation. γ-Cyclodextrin does not appear to be as selective as the controlled layered spacing of the $M[N]_4X_2$ complexes which can be tailored to separate geometric isomers.

SEPARATION CAPABILITIES

The urea and thiourea systems have been thoroughly discussed by previous authors and need no repetition here regarding their classic separation behavior.

However, in recent years the urea channel has been used as a polymerization template, a release agent for selected chemicals in a reaction, and the separation of cis-trans isomers. Specificity for unsaturation position has also been noted.

The shorter the chain length required, the greater the ease of urea complex formation. Dialkyl esters of succinic, fumaric, maleic, and acetylenedicarboxylic acid have been studied in this regard (9).

The shortest homologs capable of complex formation in the succinate and fumarate series are the diethyl esters which have the "slenderest" cross sections (5.0 and 5.1 Å).

The acetylenedicarboxylate, which has a 5.2 Å cross section, requires the di-n-propyl ester as a minimum while the 5.3 Å diameter maleate requires the di-n-butyl ester as a minimum for complex formation (10).

Further studies in this area indicated the influence of unsaturation position on complex-forming capabilities.

Alkenoic acids, except 3-n-hexenoic and 3-n-octenoic acids, formed urea inclusion compounds.

All other alkenoic acids differed from these two by the presence of a double bond in the 2-position. The same authors noted a similar trend in studies of alkynes. The triple bond did not prevent complex formation in 1-, 2-, or 3-nonyne and 2- and 3-decyne but did prevent complex formation in 4-nonyne and 5-decyne. Thus, as these authors observed, the isomer with the more centrally located double bond appears less stable and forms less complex or no complex under comparable conditions.

Selenourea exhibits only a small difference in channel diameter from that of urea or thiourea but appears to be more selective with geometric isomers (11).

Apparently selenourea is more amenable to adoption to the shape and size of the included molecules. Selenourea seems to be much more selective, however, in its choice of guest components. An example cited noted that trans-1-t-butyl-4-neopentylcyclohexane formed an adduct but the cis isomer was not included. Thiourea was not selective in this case, which suggests the possible use of selenourea for separating isomeric alkyl naphthenes, alkyl aromatics, and the like.

Clathrates offer an interesting challenge in this period of energy consciousness. By the proper choice of agent and reaction conditions it is possible to recover desired compounds from mixtures with small temperature changes and at a relatively high loading of the clathrate compared to loadings in conventional extraction processes. As pointed out by this author and his associates, hydrates offer some promise in this area (12).

Werner complexes are also promising candidates in this regard, but in most cases the reaction rate needs considerable improvement (e.g., by one or two orders of magnitude).

ANALYTICAL AND PREPARATIVE TECHNIQUES

Urea and thiourea have been successfully exploited as separation tools for analytical separations as well as for specialty industrial products.

Although subjected to extensive study in-depth for a quarter of a century, the urea–thiourea field continues to receive attention.

A number of patents have appeared in the United States and abroad during the past year directed toward improving the mechanics of handling the urea adduction process in the petroleum industry.

However, other available clathrating agents have not been as thoroughly studied and applied. For example, the use of hydroquinone complexes for separating and storing gases has had only limited attention. The same chromatographic techniques developed for both gas and liquid separations have received only cursory examination in the clathrate field.

If one considers developed techniques for enhancing the analytical capabilities of gas chromatography via the use of added salts to the substrate, the extrapolation of various clathrating agents such as the Werner complexes, dextrins, hydroquinone, and tri-o-thymotide is apparent.

Early work in this area involved successful application of partition and adsorption chromatography for separation of mixtures differing in type and number of functional groups or for separation of members of a homologous series (13).

The columns of Kemula and Sybilska consisted of β- and γ-picoline and 2,6-lutidine nickel dithiocyanate complexes. They used concentrated solutions of potassium thiocyanate or the γ-picoline containing some KSCM in aqueous solution as a mobile phase.

More recently (14), Werner complexes have been investigated as stationary phases in gas chromatography to determine the influence of various substituents in the aromatic rings of these complexes on the elution order of molecules of varying shapes and structures.

The spirochromans have been used in adsorption chromatography separations (15). A number of spirochroman derivatives were investigated with respect to the substituent effect on the resolution of iso- and n-alkanes. The spirochromans were added to an appropriate column followed by addition of the hydrocarbon charge with subsequent collection of effluent-containing resolved species.

One can visualize the broad spectrum of potential application of various clathrating agents in analytical and preparative separation considering the choice of structures available and the wide variety of essentially inert substrates that is available.

Werner complexes dispersed in polyolefin oxide, polyglygols, polyphenyls, and the like could be used for gas chromatography work, or they could be used either in massive form on dispersed or coated on an inert support.

The same techniques could be applied to the other clathrating compounds described, except the hydrates. Most of the hydrates are particularly applicable to lower temperature and/or high pressure operation. A few Freon hydrates are functional at atmospheric pressure and from 0 to about $+10°C$.

There are exceptions. The hydrates of trialkyl sulfonium salts and tetraalkylammonium salts are similar to the gas hydrates, but the hydrate with the formula $2[i\text{-}C_5H_{11}]_4N^+F^- \cdot 76H_2O$ is a surprising substance consisting of 68% water that remains an ice until heated to 88°F.

A series of chromatographic columns can be employed to separate complex mixtures into normal and iso-structures as well as ortho-, para-, and meta-substituted aromatic species.

Considerable potential latitude is available to the ingenious investigator in both analytical and preparative chemical separations.

SEPARATIONS EXPECTED

In the classic urea or thiourea cases, one can devise multistage extraction or adsorption columns capable of yielding 99 + % purities in two to five equilibria stages depending on the complexity of the feed and the required recovery of the desired specie. In fact, this has been done for specialty hydrocarbons, particularly those of high molecular weight where desorption of zeolite molecular sieves is less practical.

The application of urea adduction for separating polar compounds has been less active. As pointed out by Bergmann (16), alkyl halides may be complexed with much shorter chain lengths than hydrocarbons, presumably due to the polarizable functional group.

Werner complexes are particularly useful for separating aromatic hydrocarbons and their derivatives.

To separate ortho isomers, one has quite an effective tool with a nickel dithiocyanate–tetramethyl benzylamine complex as shown in Table 4 (6).

The separations shown are for a single stage only. It is apparent that relatively few equilibria stages would be required to attain high purity and recovery of most species. The chlorotoluenes are relatively unimpressive but the dichlorobenzenes are readily resolved.

TABLE 4

Separation of Ortho-Isomers: [Ni(SCN)₂(α-methylbenzyl amine)₄] (6)

	Ortho-isomer in		Wt-% guest in clathrate
Compound	Feed	Clathrate	
Ethyl toluenes	33.5	74.3	13.9
Cymenes	41.0	91.7	10.8
Ethyl isopropyl benzenes	29.5	90.3	9.6
Cyclohexyl toluenes	28.6	85.8	15.7
o-Ethyl toluene and mesitylene	50.2	98.2	13.9
Dichlorobenzenes	50.3	85.4	14.2
Chlorotoluenes	35.0	45.5	13.1

TABLE 5

Separation of Para-Isomers: Tetra-(4-methyl pyridino)-nickel dithiocyanate (5)

| Compound | Para-isomer in | | % Recovery |
	Feed	Clathrate	
Chlorotoluenes	50.2	91.4	35
Dichlorobenzenes	52.8	93.0	34.8
Toluidines	49.1	68.4	10.1
Nitrotoluenes	50.0	73.5	7.0
Methyl anisoles	52.2	94.4	25.5

If one is concerned with the separation and/or recovery of para isomers, a change in the base of the complex suffices as shown in Table 5. Here is an excellent example of changing clathrating capabilities by apparent altering of the distance between the plates of the nickel dithiocyanate (5).

Para isomer separation was also studied by Williams (7) using the metal salt as a variable with surprising results as shown in Table 6. Here is a classic analogy of the nickel ammonium thiocyanate–benzene complex case wherein the 4-methyl pyridine molecule provides a large "box" or cage in the layers compared to the NH_3 case.

A comparison of nickel salt effects on dichlorobenzenes and methyl styrenes indicates excellent agreement on selectivity. The capacity of the complex for methyl styrenes is greatly improved by using the ferrous or cobaltous salts in place of the nickel. Selectivity is somewhat reduced, but this difference would be relatively insignificant in a multistage process such as liquid phase chromatography.

Much more needs to be done to exploit this interesting field. With our present knowledge, almost any desired aromatic isomer may be selectively trapped by the proper Werner complex.

As an example of altering clathrate structures to accommodate a

TABLE 6

Separation of Para-Isomers: 4-Methyl pyridine $(SCN)_2$ Clathrates (7)

Metal	Feed	Mole guest per mole metal complex	Composition of guest
Ni^{2+}	Dichlorobenzenes	0.76	90.5%p, 9.5%
Ni^{2+}	Methyl styrenes	0.60	90.0%p, 10.0%
Fe^{2+}	Methyl styrenes	0.94	82.3%p, 17.7%
CO^{2+}	Methyl styrenes	0.86	82.5%p, 17.5%

particular compound, Morita et al. (*17*) described tetracyanopalladate(II) and tetracyanoplatinate(II) aromatic clathrates. Thus $Cd[NH_3]_2Pd[CN]_4$ complex can trap fluorobenzene which apparently has too large a molecular volume to form a clathrate with the lattice $Ni[NH_3]_2Ni[CN]_4$.

The fluorobenzene clathrate $Cd[NH_3]_2Pd[CN]_4 \cdot nC_6H_5$ ($n < 2$) was confirmed by IR spectra, the broad line NMR spectra of ^{19}F, and proton.

Hydrates have been discussed by this author and his co-workers (*12, 18*) previously in some detail regarding their separation capabilities.

They are particularly useful for separating C_3 to C_5 molecular weight hydrocarbons. Excellent separations of normal and isoalkanes and alkenes have been demonstrated with better separations of paraffins than of the corresponding olefin.

In Table 7 a few of the clathrating agents that have been found to be particularly useful for low molecular weight hydrocarbon separations are tabulated.

The capabilities of the tetrahydrofuran system for separating isobutene

TABLE 7

Properties of Hydrating Agents Studied

Name	Structure	bp (°C)	Critical hydrate properties	
			T (°C)	Pressure (Torrs)
Freon, F-11	CCl_3F	+25	+6.5	410
Freon, F-21	$CHCl_2F$	+8.9	+8.7	760
Freon, F-1426	CH_3CFCl_2	−9.6	+13.1	1743
Methyl bromide	CH_3Br	+3.56	+14.7	1151
Tetrahydrofuran		+64–66		

TABLE 8

C_4 Paraffin, Olefin, Diene System

Component, wt-%	Feed	THF, hydrate
Isobutane	3.1	8.2
n-Butane	3.8	5.1
Butene-1	13.4	5.8
Isobutene	40.3	71.4
cis,trans-Butene-2	12.0	2.8
Butadiene	27.5	6.7

and isobutane from a mixture of the other C_4 hydrocarbons normally found in a cracked gas stream are illustrated in Table 8. These data have been previously reported in the above references but are offered here for review. Again it should be pointed out that these results are for single stage separations.

However, the limitations encountered have been largely attributable to the limitations of the void space in the hydrate structure. Altering the lattice structure of the hydrate via other lattice-forming hydrating agents could be the answer.

THE FUTURE OF CLATHRATE CHEMISTRY

Since the discovery of the diamine nickel(II) tetracyanoniccolate(II) dibenzene $(Ni[NH_3]_3Ni[CN]_4 \cdot 2C_6H_6)$ by Hoffmann (1) in 1897, the field of clathrate chemistry has broadened to include a wide spectrum of organic and inorganic compounds capable of high selectivity and capacity.

The opportunities to tailor a "host" to accommodate a specific guest are apparent. The potential application to analytical and preparative problems are numerous.

In many cases the use of clathrates to achieve a separation may involve relatively small temperature and/or pressure changes. Compared to extractive distillation or liquid–liquid extraction, clathrates can be relatively low in energy requirements to achieve a given separation.

As the cost of energy increases and provides a larger lever in process economics, engineering requirements will be developed to further exploit these useful and interesting compounds beyond the analytical and preparative scope emphasized here.

Acknowledgments

I wish to thank Professor Eli Grushka for inviting me to make this presentation, and Monsanto Polymers and Petrochemicals Co. for permission to do so.

REFERENCES

1. R. A. Hofmann and F. Kuspert, *Z. Anorg. Allg. Chem., 15*, 204 (1897).
2. M. F. Bergen, German Patent Appl. OZ123438 (March 18, 1940).
3. A Clemm, *J. Liebigs Ann. Chem., 110*, 357 (1859).
4. H. M. Powell and J. H. Rayner, *J. Chem. Soc., 1952*, 319–328.
5. W. D. Schaeffer et al., *J. Amer. Chem. Soc., 79*, 5870 (1957).

6. P. de Radzitsky and J. Hanotier, *Ind. Eng. Chem., Process Des. Develop.* *1*(1), 10 (1962).
7. F. V. Williams, *J. Amer. Chem. Soc.,* *79*, 5876 (1957).
8. H. R. Allcock and L. A. Siegel, *Ibid.,* *86*, 5140 (1964).
9. J. Radell, B. W. Brodman, and E. D. Bergmann, *Tetrahedron,* *20*(1), 13 (1964).
10. J. Radell and B. W. Brodman, *Can. J. Chem.,* *43*, 304 (1965).
11. H. van Bekkum, D. J. Remijonse and B. M. Wepster, Chem. *Commun.,* *2,* 67 (January 22, 1969).
12. H. J. Gebhart et al., *Chem. Eng. Progr. Symp. Ser.,* *66*(103), 105–115 (1970).
13. W. Kemula and D. Sybilska, *Nature,* *185*, 237 (January 23, 1960).
14. A. C. Bhattacharyya and A. Bhattacharje, *Anal. Chem.,* *41*(14), 2055 (1969).
15. Edw. M. Geiser, U.S. Patent 2,851,500 (September 9, 1958), to Universal Oil Products.
16. E. D. Bergmann, *Can. J. Chem.,* *42*, 1069 (1964).
17. M. Morita et al., *Bull. Chem. Soc., Japan,* *40*(6), 1556 (1967).
18. E. S. Perry and C. J. van Oss, eds., *Separation and Purification Methods,* Vol. 1, Dekker, New York, 1973, pp. 371–407.

Potential-Controlled Adsorption at Chemically Modified Graphite

J. H. STROHL and K. SEXTON

DEPARTMENT OF CHEMISTRY
WEST VIRGINIA UNIVERSITY
MORGANTOWN, WEST VIRGINIA 26506

Abstract

The use of some chemically modified graphites for separation is described. Chemically bonded sulfonic acid groups can apparently be added by the use of fuming sulfuric acid.

INTRODUCTION

Potential dependent adsorption of compounds at an electrode surface is a fairly commonly observed process and has been studied by a variety of techniques. The use of this process for separation and removal has only recently been studied. Packed bed electrodes which have a large surface area and efficient solution contact have made the study of this process practical.

The reasons for studying this process are:

(1)   Any new method may have application to the separation of mixtures that are presently difficult.

(2)   It may be capable of rapid batch removal of single components.

(3)   It is a rapidly reversible process since only the potential has to be changed, making the process suitable for regenerative removal and separation systems.

Potential controlled adsorption has been applied to the separation of mixtures of dyes (*1*), to the separation of mixtures of quinones, and to the concentration of dilute solutions of quinones (*2*).

EQUIPMENT

The cells necessary for these experiments are relatively simple to construct and maintain. The essential parts are:

(1)   A packed electrode bed of small enough particle size that efficient, rapid contact can be achieved. The sample solutions are passed through this bed.

(2)   Some sort of a porous barrier which has a relatively high solution conductivity. This barrier serves to prevent gross mixing of the sample solution with counter electrode solution but allows for ionic conduction.

(3)   A counter-reference electrode compartment which is necessary for the control of the potential of the packed electrode bed.

The cell should have relatively uniform, low-resistance paths between electrodes to minimize potential inhomogenieties on the packed electrode bed. Relatively low currents are required for potential controlled adsorption experiments and therefore relatively low potential (IR) gradients are produced so that cell design is not as critical as for high current applications. Many cell designs have been published that appear to be suitable for this application, although most were not used for potential controlled adsorption experiments. The most recent design (*3*) used a silver packed bed electrode and gives reference to other suitable designs.

Since these experiments are performed at essentially constant potential and the currents are low, any standard potentiostat circuit may be used (even manual). A pump is convenient for the control of flow rate although gravity flow has been used successfully. Some type of detection system is necessary for the observation of amounts; the type would be determined by the nature of the compounds being studied.

ELECTRODE MATERIALS

The electrode materials or adsorbents must be conducting, relatively unreactive in the potential range to be used, and available in a form suitable for a packed bed. Many metals and some forms of carbon are suitable. In addition, modified forms such as conductors with adsorbed or chemically bonded surface layers may have interesting separation properties.

At present a relatively nonporous graphite is the only adsorbent that has been extensively studied (*1, 2*). It appears to have a high affinity for aromatic compounds and very low affinity for metal ions. The capacity of this adsorbent for most quinones is in the range of 1 to 5 μmoles/g of graphite (105 to 147 μ size). Other, more porous graphites and conducting carbons would probably be useful as adsorbents and might have higher capacities. Platinum and perhaps mercury-coated platinum would be useful. There have been numerous observations of potential controlled adsorption at these surfaces although they have not been used for separations. Other less noble metals may be useful in specific cases, but would probably not be as generally useful.

The modified forms appear to be potentially useful in the area of selective removal by tailoring the surface layer to suit the problem. A recent report by Lane and Hubbard (*4*) serves as an excellent example. They adsorbed 3-allylsalicylic acid on a platinum electrode in a thin-layer cell and observed the accumulation of ferric ion at the adsorbed layer, apparently by chelation. When the potential of the electrode was changed to a more positive value, the accumulation of ferric ion was negligible. From a simple calculation based on surface area and the amount of 3-allylsalicylic acid adsorbed per unit area, a bed of 40 μ platinum particles should have a capacity of about 0.1 to 1 μmole of ferric ion per cubic centimeter of packing material. This would compare favorably with the capacities exhibited by graphite and should have much more selective properties. The adsorption of other olefins with similar selective properties is also described and should have utility for other separations.

Two recent studies in this laboratory on modified graphite surfaces indicate that graphite is also suitable for the production of selective adsorbents for potential controlled adsorption. Adsorption of the sodium salts of 2-naphthalenesulfonic acid and 2,7-naphthalenedisulfonic acid on graphite produces a material with ion-exchange properties. These can be converted to the acid form and then the ion-exchange capacity can be determined by passing a sodium chloride solution through and determining the liberated hydrogen ion by titration. These materials exhibit exchange capacities of about 2 μmoles/cc of graphite, which is in reasonable agreement with the known adsorption capacities for the sulfonic acids.

The second study was the production of a surface which is apparently composed of chemically bonded sulfonic acid groups. Graphite treated with fuming sulfuric acid (30% SO_3) produces a material which has an exchange capacity of 200 μmoles/cc initially and decreases to about 80 μmoles/cc after 20 days as determined by the hydrogen ion liberated by sodium chloride. Preliminary experiments with ferric ion in hydrochloric and in perchloric acid media indicate that it can be adsorbed at some potentials and desorbed at other potentials. For example, in 0.05 M perchloric acid the retention volume for ferric ion at $+0.48$ V (vs SCE) is over twice the retention volume at $+0.71$ V.

MODES OF USE

Potential-controlled adsorption at graphite has been applied to two types of problems; concentrating dilute samples and the single stage separation of mixtures. It has not been applied to a chromatographic separation although this should be possible for systems with rapid adsorption kinetics.

The concentrating of dilute samples is accomplished by applying a potential at which the compound of interest is adsorbed, passing a large volume of a dilute solution through the packed bed electrode, and then changing the potential so that the accumulated compound is released to a small volume of solution. Concentrating factors of as high as 50 have been achieved. Attempts to concentrate samples further are limited by less than quantitative recovery, apparently due to migration of sample through the bed. If solution conditions are chosen so that the extent of adsorption is higher, then the desorption step may become more difficult. This mode is very similar to the plating-stripping technique for concentrating metals.

The single stage separation of mixtures is based on differences in the potential-adsorption properties for different components of the mixture. For example, consider a two-component mixture. If a potential exists such that one component is adsorbed and the other is not, mixtures of the two can be separated because one component would be held on the packed bed electrode while the other component passed through. More complex mixtures would require a potential such that all but one component is adsorbed, a second potential such that all but two components are adsorbed, etc. However, due to the migration effect, those separations would probably be less than quantitative.

The application of modified electrode surfaces to the above types of problems might lead to improvements in concentrating factors and degree of separation because of much greater specificity. The general method appears to have application to regenerative removal systems for specific materials because of the ease and low cost of the regeneration step. A general example would be the removal of a specific pollutant from mine drainage or an industrial discharge.

REFERENCES

1. W. B. Caldwell and J. H. Strohl, Department of Chemistry, West Virginia University, Morgantown, West Virginia, Unpublished Work, 1972.
2. J. H. Strohl and K. L. Dunlap, *Anal. Chem.*, *44*, 2166 (1972).
3. R. E. Meyer and F. A. Posey, *J. Electroanal. Chem.*, *49*, 377 (1974).
4. R. F. Lane and A. T. Hubbard, *J. Phys. Chem* , *77*, 1401 (1973).

Electrokinetic Cell Separation

CAREL J. VAN OSS

DEPARTMENT OF MICROBIOLOGY
STATE UNIVERSITY OF NEW YORK AT BUFFALO
BUFFALO, NEW YORK 14214

Abstract

Particles of the size of living cells are too big to be separated in stabilizing media such as gels. The separation thus has to be done while they are freely suspended in a buffer contained in a vessel of appropriate shape. The major two disturbing factors under these conditions are cell sedimentation and electroosmotic backflow along the walls of the vessel.

Cell sedimentation can be overcome (a) by measuring very small cell transport paths, along which the cells have no time to sediment, using a microscope; (b) by applying a continuous liquid flow perpendicular to the direction of separation; (c) by stabilizing the cells in a density gradient; (d) by ascending electrophoresis, in the direction opposite to that of gravity; (e) by using a column packed with beads of a homogeneous size; and (f) by doing the separation at zero gravity, in outer space.

Electroosmotic backflow. This can (a) be overcome by coating the vessel-walls with a material of 0 ζ potential; and (b) be managed by forcing it into a completely uniform flow by coating the vessel walls with a material of non zero ζ-potential and by providing both ends of the vessel with porous plugs made of the *same* material.

INTRODUCTION

Only the electrophoresis of the largest units that can undergo this process to any useful purpose has to contend with the detrimental action of both sedimentation and electroosmosis.

All particulate and soluble materials of a dimension smaller than a few hundred nanometers can be stabilized by gels (*1*). The use of gels or

71

porous blocks obviates all sedimentation and most convection problems, while the proper choice of gel pore size can generally eliminate the retardation of particles or molecules due to the exiguity of the gel's pores (2). Electroosmotic backflow is, of course, by no means eliminated by the use of gels or porous blocks, but the electroosmotic flow in gels or blocks is completely uniform and is thus of small importance in preparative separations, while it can easily be quantitatively taken into account in analytical electrophoretic determinations.

Thus only those particles that are too large to be stabilized by gels are subject to the twin disturbing actions of sedimentation and electroosmosis. The most important of such particles in the field of biology undoubtedly are living cells. In microscopic electrophoresis of living cells, sedimentation is no problem, due to the fact that cells need only be observed for a few seconds, during which they migrate only a few micrometers and have no opportunity to sediment to any extent. In the classic microelectrophoretic method, however, electroosmotic backflow along the walls of the capillary is the great drawback. On the other hand, in preparative cell electrophoresis, although electroosmosis may play a certain role, the major impediment is the strong tendency of all cells to sediment to the bottom of the electrophoresis vessel within the time required for the separation.

SEDIMENTATION

Microelectrophoresis

As already mentioned above, cell electrophoresis during very brief time spans over very short pathways does not give rise to a noticeable sedimentation. The great problem here is the electoosmotic backflow, which is treated below.

Continuous Flow Electrophoresis

Continuous flow electrophoresis is a fairly complicated process in which cells are suspended in a steadily flowing stream of liquid, perpendicular to which an electric field is applied. When all parameters are well controlled, a steady state can be reached, under which the cells separated by the electric field while being transported in the flowing liquid vein, continuously arrive at the same exit ports, according to their particular electrophoretic mobility. Electroosmosis plays a considerable role in this method, which makes it only possible to focus sharply one fraction at a time (see below). Nevertheless, by this method some important cell separations have already been accomplished (3–7). Endless belt electrophoresis, developed by Kolin (8), is another variant of continuous flow electrophoresis.

Stabilization in a Density Gradient

It is, of course, theoretically possible to levitate any object heavier than water by adding solutes to the water until a density is attained that is close to that of the object to be levitated. In practice, a density *gradient* rather than a uniform high density is more generally utilized, because such a gradient also has a stabilizing power against minor thermal and other convections. Nevertheless, and in particular for cells, which need stabilization most, the presence of high concentrations of solute in the water unfortunately also tends to give rise to strongly hypertonic osmotic pressures. In order to obviate this, high-molecular-weight materials have been proposed for the elaboration of density gradients, e.g., by Boltz et al. (9), who after thorough analysis of the available macromolecules have chosen a cross-linked dextran of molecular weight 400,000 for the purpose (Ficoll, Pharmacia, Piscataway, New Jersey). For some cells, however, such high-molecular-weight polymers cannot be used, as they give rise to severe cell aggregation. However, we have found that lymphocytes, which are among the more important cell types that are in need of electrophoretic separation, will not aggregate when the lowest available molecular weight of dextran is used, i.e., a linear dextran of a molecular weight of 10,000 (Dextran T–10, Pharmacia, Piscataway, New Jersey) (10). Nevertheless, even with the lower molecular weight of polymers, their concentration in a density gradient is generally such that a considerable amount of polymer adsorption onto the cell surface must be feared, giving rise to a change in ζ-potential which, of course, complicates the intended electrophoretic separation.

Ascending Electrophoresis

We recently observed that in the total absence of a density gradient, ascending electrophoresis of cells can be practiced in vertical tubes (10). By this method it has proved possible to separate human lymphocytes into three groups of different electrophoretic mobilities. Upon once or twice repeated reelectrophoresis of the fastest group, virtually pure T cells could be obtained in quantities of the order of 10^7 cells. Isolation of the electrophoretically slower B cells is somewhat more complicated because the slowest moving fraction is also the one that remains most contaminated by the few cells that form aggregates and thus are heavier and tend to remain close to the bottom of the tube. Fortunately, B cells are not only the slowest ones electrophoretically speaking but also have the greatest tendency to aggregate. Nevertheless, upon aggregation some T cells apparently remain entrapped. The electrophoretic purification of B cells therefore is strongly linked to the problem of complete avoidance of cell aggregation.

Columns Packed with Beads of a Homogeneous Size

To separate cells that are all of virtually identical size but that consist of groups of different electrophoretic mobilities, such as erythrocytes, tubes packed with glass beads of homogeneous size can be used with advantage. To prevent the cells from sticking to the glass beads, the beads have to be treated with polyoxyethylene. In order to avoid the large interstices that tend to occur between the glass beads and the wall of the tube, the wall of the tube should be coated with a gel (agarose, for instance) in which the glass beads, after being packed into the column, can partly imbed themselves. An additional advantage of such an inner gel coating is that after the electrophoresis the entire column can be extruded and sliced in order to obtain the various fractions. We found that the optimal diameter of the glass beads for the separation of human erythrocytes is approximately 100 μm (or 100–140 mesh). In a homogeneously packed tube the diameter of the smallest spherical particle that can just be retained in the beads' interstitial spaces is approximately $15\frac{1}{2}\%$ of the diameter of the glass beads themselves (11). However, for cells that occur in families of various sizes, this method cannot be used, as the larger cells will remain entrapped in the interstitial spaces while the smaller cells will too readily be able to sediment unhindered through these spaces.

Electrophoresis at Zero Gravity in Outer Space

When gravity can be entirely abolished, sedimentation no longer occurs. That condition can only be maintained for any length of time in outer space. For that reason a series of experiments has been done, and others are in advanced stages of preparation, for electrophoresing cells in outer space. An experiment in Apollo 16 with two different sorts of latex particles showed that sedimentation was indeed entirely abolished at zero gravity

(*12*), although electroosmosis still was an important problem in that experiment. More recently the electrophoretic separation of red blood cells by isotachophoresis has been attempted in Skylab 3 (*13*), while further electrophoretic separations of erythrocytes as well as of lymphocytes and other cells are projected for the impending Apollo-Soyuz flight.

ELECTROOSMOSIS

Zero ζ-Potential Coatings

Glass walls generally have a considerable ζ-potential. But there exist various techniques for covalently binding a variety of compounds of much reduced or even of zero ζ-potential to glass (*14*). In particular, the use of proteins of the appropriate isoelectric point seems promising. Proteins can be covalently bound to glass (*15*). A treatment for agarose has recently been published that reduces its ζ-potential to zero (*16*). With such treated agarose gels, glass tubes may be coated and in that fashion their electroosmosis can be abolished. Tubes made of various plastics frequently will be found to have a much lower ζ-potential than glass and can often be used without any further coating or precaution. When, however, material such as protein is present in solution together with the cells, it must be kept in mind that most plastics quite readily adsorb protein (*17*) and many other materials and will thus acquire the ζ-potential of the material that they adsorb. However, in a protein-free medium, plastics as construction material for electrophoresis tubes may be quite satisfactory, at least for preparative purposes.

Uniformization of the Electroosmotic Backflow

In microelectrophoresis with closed capillaries the particular drawback of electroosmosis is that it is very strong along the side of the wall of the capillary, so that the mobilities found for cells or particles differ tremendously according to the exact place inside the capillary. Numerous electrophoretic mobilities of particles must therefore be plotted as a function of the depth of focusing in the capillary. On the parabola thus obtained, the true electrophoretic mobility can be found by interpolation (*18*). This process is not only lengthy but also gives rise to various inaccuracies.

We have succeeded in abolishing this inconvenience by coating the capillary with a gel of a given ζ-potential and by occluding both ends of the capillary with the same gel. This creates an electroosmotic backflow that is entirely uniform throughout the remaining lumen of the capillary. The magnitude of this backflow can be determined with great precision by the simple electrophoresis of an uncharged substance in a slab of the same gel. Thereupon the electrophoretic mobilities of the cells inside such capillaries can be determined at *any* level of focusing and will, after simple addition of the electroosmotic correction factor, directly yield the electrophoretic mobility (*19*). Microelectrophoresis using uniformized electroosmosis as described above can therefore be done without any special apparatus except for a microscope, a dc power supply, and a few capillary tubes. This diminishes the time it takes to determine the electrophoretic mobility of a given family of cells considerably.

In preparative electrophoresis the uniformization of electroosmotic flow can be made use of to increase the effective electrophoretic path length, in what actually can be a fairly short tube, by arranging a strong, uniform electroosmotic backflow in one direction.

Effects of Electroosmosis in Continuous Flow Electrophoresis

In continuous flow electrophoresis, electroosmotic flow perpendicular to the main direction of liquid flow causes the electrophoretically separated bands to assume crescent-shaped cross sections which, according to the electrophoretic mobilities, are either concave or convex with respect to the anode (*20*). It is, however, possible to have *one* of the electrophoretic fractions migrate as a flat band so that that fraction can be collected with optimum purity (*21*). For the precise definition of *all* electrophoretic fractions, the crescent phenomenon must be abolished, which can only be done by a zero ζ-potential coating.

BUFFERS

Zeta potentials (*22*) and thus, generally speaking (for further precision, see Refs. *23* and *24*), electrophoretic mobilities of cells and particles are higher at low ionic strength. In the same electric field the Jouleian heat development is lower at low ionic strength. There is thus every advantage in electrophoresing cells in low ionic strength buffers which, however, need to be made osmotically isotonic by the addition of nonelectrolyte solutes. One such buffer, pH 7.7 and ionic strength $\Gamma/2 = 0.01$, with an osmolality of 256 m Osm./liter and consisting of 0.18% Na_2HPO_4, 0.02% KH_2PO_4, and 4.32% glucose, can be used in most cases.

COOLING

Both in preparative and in microelectrophoresis, cooling arrangements are indispensable for the dissipation of the Jouleian heat.

In preparative electrophoresis, even at an ionic strength of $\Gamma/2 = 0.01$, a field of 5.5 V/cm necessitates a total potential difference of 500 V at 6 mA, generating 3 W. The most appropriate cooling method is by circulating cold water in a water jacket surrounding the electrophoresis tube (*10*).

In microelectrophoresis the main reason for cooling lies in the need for keeping the viscosity and dielectric constant unchanged. A constant temperature is easily attained in the microelectrophoresis method described above by placing the capillary in a trough (made from a longitudinally split plastic tube) through which water (at 20°C) from a higher placed reservoir on the left is allowed to stream horizontally and to run out to a sink on the right (*19*).

Acknowledgments

Supported in part by PHS Grant No. CA 02357 and NASA Contract No. NAS8–29745. The author is much indebted to Mr. R. M. Fike, Mr. P. M. Bronson, and Dr. P. E. Bigazzi, and Dr. C. F. Gillman for their collaboration and advice.

REFERENCES

1. F. Milgrom and U. Loza, *J. Immunol.*, **98**, 102 (1967).
2. H. Bloemendal, *Zone Electrophoresis in Blocks and Columns*, Elsevier, New York, 1963.
3. K. Zeiller and K. Hannig, *Hoppe-Seyler's Z. Physiol Chem.*, **352**, 1162 (1971).
4. K. Zeiller, K. Hannig, and G. Pascher, *Ibid.*, **352**, 1168 (1971).
5. K. Zeiller, E. Holzberg, G. Pascher, and K. Hannig, *Ibid.*, **353**, 105 (1972).
6. K. Zeiller, G. Pascher, and K. Hannig, *Prep. Biochem.*, **2**, 21 (1972).
7. K. Zeiller and G. Pascher, *Eur. J. Immunol.*, **3**, 614 (1973).
8. A. Kolin and S. J. Luner, in *Progress in Separation and Purification*, Vol. 4 (E. S. Perry and C. J. van Oss, eds.), Wiley-Interscience, New York, 1971, p. 93.
9. R. C. Boltz, P. Todd, M. J. Streibel and M. K. Louie, *Prep. Biochem.*, **3**, 383 (1973).
10. C. J. van Oss, P. E. Bigazzi, C. F. Gillman and R. E. Allen, *Proceedings of the AIAA 12th Aerospace Meeting*, Paper No. 74–211, 1974; *Proceedings of the Symposium on Processing in Space*, NASA, Huntsville, Alabama, 1974.
11. R. M. Fike and C. J. van Oss, *Prep. Biochem.*, **3**, 183 (1973).
12. R. S. Snyder, M. Bier, R. N. Griffin, A. J. Johnson, H. Leidheiser, F. J. Micale, J. W. Vanderhoff, S. Ross, and C. J. van Oss in *Separation and Purification Methods*,

Vol. 2 (E. S. Perry, C. J. van Oss, and E. Grushka, eds.), Dekker, New York, 1973, p. 259.

13. R. S. Snyder and M. Bier, *Proceedings of the AIAA 12th Aerospace Meeting*, Paper No. 74–210, 1974.

14. R. A. Messing, P. F. Weisz, and G. Baum, *J. Biomed, Mater. Res.*, *3*, 425 (1969).

15. H. Weetall, *Science*, *166*, 615 (1969); *Nature*, *232*, 474 (1971).

16. A. Grubb, *Anal Biochem.*, *55*, 582 (1973).

17. C. J. van Oss and J. M. Singer, *J. Reticuloendothelial Soc.*, *3*, 29 (1966).

18. G. V. F. Seaman, in *Cell Electrophoresis* (E. J. Ambrose, ed.), Little, Brown, Boston, 1965, p. 4.

19. C. J. van Oss, R. M. Fike, R. J. Good, and J. M. Reining, *Anal. Biochem.*, *60*, 242 (1974).

20. A. Strickler and T. Sacks, *Prep. Biochem.*, *3*, 269 (1973).

21. G. D. McCann, J. W. Vanderhoff, A. Strickler, and T. Sacks, in *Separation and Purification Methods*, Vol. 2 (E. S. Perry, C. J van Oss, and E. Grushka, eds.), Dekker, New York, 1973, p. 153.

22. A. J. Rutgers and M. de Smet, *Trans. Faraday Soc.*, *41*, 764 (1945).

23. P. H. Wiersema, A. L. Loeb, and J. Th. Overbeek, *J. Colloid Interfac. Sci.*, *22*, 78 (1966).

24. C. J. van Oss, in *Techniques of Surface and Colloid Chemistry and Physics*, Vol. 2 (R. J. Good and R. R. Stromberg, eds.), Dekker, New York, 1975.

Isoelectric Focusing: Fundamental Aspects

NICHOLAS CATSIMPOOLAS

BIOPHYSICS LABORATORY
DEPARTMENT OF NUTRITION AND FOOD SCIENCE
MASSACHUSETTS INSTITUTE OF TECHNOLOGY
CAMBRIDGE, MASSACHUSETTS 02139

Abstract

The principles, history, types, and applications of isoelectric focusing are presented. Also discussed are the theoretical aspects, electrooptical scanning, methodological parameters, and apparent physical constants. The lack of suitable carrier ampholytes is the primary reason why isoelectric focusing has not reached its full potential.

INTRODUCTION

Principle

Isoelectric focusing is an electrophoretic method which utilizes the migration behavior of amphoteric molecules in a pH gradient to achieve their condensation into narrow isoelectric zones that are stationary in the electric field. The steady-state position of each zone in the pH gradient depends on the isoelectric point (pI) of a particular amphoteric molecule, therefore isoelectric focusing can be used as a separation technique. The method involves mainly two processes which can be carried out either simultaneously or separately. These include (a) the formation of a stable pH gradient which increases from the anode to the cathode, and (b) the electrophoretic migration of the amphoteric molecules under study (e.g., proteins) toward their respective pI positions with subsequent attainment of the steady-state. At present the stable pH gradient is formed by the use of carrier ampholyte mixtures with specific properties which contain components with pI's within a defined pH range.

In order to understand how isoelectric focusing takes place, let us consider a model system where convective remixing of separating components does not occur. The anode (positive charge electrode) is placed in a strong acid (e.g., phosphoric acid) and the cathode in a strong base (e.g., NaOH). The two electrodes are then connected to a dc constant voltage power supply to produce an electric field between them. The negatively charged cathode in the electrolysis cell attracts positive ions and repels negative ions; the opposite occurs at the anode. If an amphoteric compound is present in the system, such as a carrier ampholyte, it will become negatively charged at the cathode and positively charged at the anode. This causes repulsion of the ampholyte from the electrodes. The ampholyte which has the lowest pI (most acidic) will migrate closer to the anode where it will condense in its isoelectric state (zero net charge) at some distance from the anode; the opposite will occur for the ampholyte with the highest pI (most basic) where it will condense close to the cathode. If a mixture of carrier ampholytes is used with intermediate pI values, these will focus at different positions along the electric field so that a pH gradient is formed which is defined by the pH of the ampholytes at the point of focusing. The nature of the pH gradient will depend on the range of isoelectric points, the number of carrier ampholyte species in the system, and their relative concentration and buffering capacity.

The formation of a stable pH gradient with adequate buffering capacity and conductance provides the basis for the isoelectric focusing of amphoteric molecules of interest. This process involves electrophoretic migration in a pH gradient subject to the properties of the pH-mobility curve of each particular species. The addition of an anticonvection medium (e.g., density gradient, gel matrix) and a zone detection system (e.g., UV absorbance, staining) is necessary for the practice of the method either at the preparative or analytical levels.

History

Early "stationary electrolysis" experiments involving isoelectric condensation of dyes (*1*) and glutamic acid (*2*) did not receive any attention probably because the principle of isoelectric focusing was not clearly expressed. Several years passed by until Williams and Waterman (*3*) defined the basic concept of isoelectric condensation and performed experiments with a multichambered apparatus in which adjacent compartments were separated by membranes to avoid remixing of isoelectric components. Sporadic applications of this technique for separation of biological materials including proteins was reported in the scientific literature (*4–8*). However, the method suffered from the lack of a stable pH gradient of adequate conductance and the use of a severely limited anti-convection system.

The next stage in the development of the isoelectric focusing technique was the introduction of short-lived "artificial" pH gradients by Kolin (*9–12*). These pH gradients–which were produced by mixing two buffer solutions–were not stable in the electric field because the buffer ions migrated electrophoretically. However, Kolin realized the importance of forming a pH gradient with sufficient buffering capacity and uniform high conductivity. He also introduced a density gradient system to overcome convective disturbances. Several other systems utilizing poorly reproducible and unstable pH gradients were also reported in the middle fifties (*13–16*). Svensson (*17, 18*) introduced the concept of carrier ampholytes which were defined as ampholytes with appreciable conductance and buffer capacity in the isoionic form. He realized that for isoelectric focusing to be successful, it is necessary to have several species of ampholytes with isoelectric points distributed throughout the pH range of interest. Svensson also derived the differential equations describing the dynamic equilibrium between diffusion and isoelectric condensation at the steady-state. In addition, the design of a preparative isoelectric focusing column utilizing density gradients for stabilization of protein zones was reported by him (*18*). Thus the pace was set for meaningful isoelectric focusing experiments which were not realized because of the lack of suitable compounds to serve as carrier ampholytes. However, this was remedied by Vesterberg (*19*) who was able to synthesize a mixture of polyaminopolycarboxylic acids which met the requirements for the formation of a natural pH gradient. The first successful modern isoelectric focusing experiment was reported by Vesterberg and Svensson (*20*) which opened the way to the widespread use of the method with the help of the LKB Company which made commercially available Svensson's preparative column and Vesterberg's carrier ampholytes. Shortly after Vesterberg's and Svensson's (*20*) publication, a number of microtechniques utilizing gels as stabilization medium were reported in the literature (*21–32*). The further development of new gel and density gradient microcolumn methods [for reviews, see Catsimpoolas (*33*), (*34*)] contributed to the popularization of the method because of the advantages of rapid analysis, low cost, and widely diversified selection of detection methods.

Recently, Catsimpoolas and co-workers (*35–42*) reported the development of the "transient state isoelectric focusing" (TRANSIF) method based on a new kinetic theory and the use of an analytical instrumentation system utilizing continuous *in situ* electrooptical monitoring of the separation path (*43*). This new technique made possible the direct on-line measurement of methodological parameters and apparent physical constants, and the evaluation of the "steady-state" of isoelectric focusing. It was soon realized that TRANSIF offered the only feasible approach to the exact evaluation of an isoelectric separation because of instabilities in the pH gradient and the presence of nonuniform conductance along the separation field.

Types of Isoelectric Focusing

There are several ways to classify isoelectric focusing experiments in some useful operational fashion (*33, 34*). The distinction between preparative and analytical isoelectric focusing pertains to the end result of either collecting the separated material at the conclusion of the experiment, or being interested only in the analytical aspects without consideration for the fate of the sample. Usually preparative isoelectric focusing is carried out in large columns (100 to 400 ml volume) using a sucrose density gradient as supporting medium. Alternatively, blocks of granular gels (e.g., Sephadex) or zigzag horizontal electrolysis cells can be used. Proteins are isolated in milligram or even gram quantities. On the other hand, analytical techniques can be performed with microgram or even nanogram amounts of material in small columns (1 to 10 ml) of density gradient or gel, and in flat gel slabs.

In regard to the kind of supporting medium, or the absence of it, we can distinguish several types of isoelectric focusing experiments. These include *density gradient, gel, zone convection*, and *free solution* isoelectric focusing. The density gradient is formed using two different concentrations of sucrose or other neutral substances such as Ficoll and ethylene glycol with special devices, or by layering fractions of different density. Gel isoelectric focusing is carried out either in homogeneous (continuous phase) or granular (beads) polyacrylamide, agarose, or cross-linked dextran gels. Zone convection and free solution isoelectric focusing requires specially constructed apparatus.

The application of specific detection methods involving antigen–antibody precipitin reactions in gels after isoelectric separation gave rise to the technique of *immunoisoelectrofocusing*. Another detection method utilizing UV absorption optics for *in situ* continuous electrooptical monitoring of the separation field has been called *scanning* isoelectric focusing. Finally, if the kinetic aspects of pH gradient electrophoresis focusing and isoelectric diffusion can be measured by electrooptical methods, the technique is called *transient state* isoelectric focusing.

Applications

The main applications of isoelectric focusing to-date have been in the field of protein and peptide separation and characterization at both the preparative and analytical levels. Proteins from various sources (e.g., animal, plant, and microbial origin) having different functions, such as enzymes, hormones, immunoglobulins, and toxins, have been isolated in homogeneous form by isoelectric focusing. Usually the method is applied to the final step of purification preceded by other fractionation techniques such as ammonium sulfate precipitation, gel filtration, and ion-exchange chromatography. Often a protein that has been found homogeneous by other methods (e.g., ultracentrifugation, chromatography, electrophoresis) can be resolved into several components by isoelectric focusing. This occurs because of the high resolving power of the technique being able to separate proteins differing by only a few hundredths of a pH unit in their isoelectric points. The observed microheterogeneity may be due to: (a) minute differences in the primary structure (asparagine and glutamine residues are part of the primary structure); (b) conformational isomers; (c) denaturation; (d) presence of variable moieties such as carbohydrates, lipids, and metals which alter the isoelectric point; and (e) strong complexing with ionic and nonionic compounds including the carrier ampholytes themselves.

Direct determination of the isoelectric point of a protein is another great advantage of the isoelectric focusing technique, and this extra feature is utilized in most applications involving separation of a mixture of proteins. Measurement of the isoelectric point is one of the physicochemical parameters required for the characterization of an unknown protein, or establishing the identity of two similar proteins.

In addition to the above applications, isoelectric focusing can be used in two-dimensional separations in gels in combination either with other electrophoretic techniques or with immunodiffusion. Such methods are very useful in resolving protein components in a complex mixture and have very important applications in fields such as clinical chemistry, biochemical genetics, and plant taxonomy.

THEORETICAL ASPECTS

The basic theory of isoelectric focusing dealing with the "steady-state" was developed by Svensson (*17*, *18*) and with the "transient-state" (kinetic) by Weiss, Catsimpoolas, and Rodbard (*40*). Knowledge of the basic assumptions, physical parameters, and equations involved in the present state of the theory of isoelectric focusing is important in the interpretation of results obtained with the technique and in the further advancement of the method both from the theoretical and methodological point of view.

Isoelectric and Isoionic Points

The isoelectric point of a protein is defined as the pH at which an applied continuous electric field (dc) has no effect on the electrophoretic migration of the molecule in respect to the solvent (*44*). In the absence of complexing ions other than protons, the isoelectric point of an amphoteric molecule is virtually identical to its isoionic (isoprotic) point (*44–47*). Knowledge of the isoionic (isoprotic) point of a protein is very valuable because it is characteristic of its intrinsic acidity. The isoionic point of a protein solution is defined as that pH which does not change when a small amount of pure protein is added to the solution. In this regard the isoelectric point of a protein determined by isoelectric focusing should be very close to its isoionic point if the following requirements are met: (a) the carrier ampholytes completely dominate the buffering of the isoelectric zone site; that is, increasing amounts of focused protein has no effect on the pH; (b) the carrier ampholytes do not complex with the protein; and (c) ions that can form complexes with the proteins are removed either by migration toward the electrodes or by strong binding to the carrier ampholytes.

Direct comparison of isoelectric points measured by isoelectric focusing with those determined by other electrophoretic methods often shows that higher pI values are obtained with the former technique (*20*). This phenomena occurs because the buffer ions used in electrophoresis can form complexes with the protein thus altering its isoelectric point. Generally, the lower the ionic strength of the buffer, the higher are the pI values. Extrapolation of the pI values determined by electrophoresis to zero ionic strength should produce isoelectric points approaching those obtained by isolectric focusing.

Dissociation Theory

This theory is given here essentially as presented by Rilbe (*48*).

Dissociation of Biprotic Ampholytes

If we consider an ampholyte consisting of its zwitterionic form HA, its anion A^-, and its cation H_2A^+, the two mutual chemical equilibria of the three subspecies are:

$$HA + H^+ \rightleftarrows H_2A^+ \tag{1}$$

$$HA \rightleftarrows A^- + H^+ \tag{2}$$

giving rise to the mass–action equations

$$K_1 = \frac{[HA][H^+]}{[H_2A^+]} \tag{3}$$

$$K_2 = \frac{[A^-][H^+]}{[HA]} \tag{4}$$

where K_1 and K_2 are the thermodynamic equation constants and [ ] denotes activity. Introducing the simpler notation:

h = $[H^+]$
pH = $-\log h$
C_+ = concentration of cation H_2A^+
C_- = concentration of anion A^-
C_0 = concentration of molecular and zwitterionic ampholyte
C = $C_+ + C_0 + C_-$ = total concentration of ampholyte
f_+ = activity coefficient of cation
f_- = activity coefficient of anion

We can write instead of Eqs. (3) and (4):

$$hC_0 = f + K_1C_+ \tag{5}$$

$$hC_- = C_0(K_2/f_-) \tag{6}$$

Activity coefficient for the uncharged species is assumed to be unitary. For simplification, the stoichiometric dissociation constants K'_1 and K'_2 are introduced, such that

$$f_+K_1 = K'_1 \tag{7}$$

$$K_2/f_- = K'_2 \tag{8}$$

Equations (5) and (6) become

$$C_+ = hC_0/K'_1 \tag{9}$$

$$C_- = K'_2 C_0/h \tag{10}$$

by adding them together with C_0 to obtain the total concentration, one arrives at a set of equations describing the concentration of the three subspecies:

$$C_+ = h^2 C/(h^2 + hK'_1 + K'_1 K'_2) \tag{11}$$

$$C_0 = hK'_1 C/(h^2 + hK'_1 + K'_1 K'_2) \tag{12}$$

$$C_- = K'_1 K'_2 C/(h^2 + hK'_1 + K'_1 K'_2) \tag{13}$$

Mean Valence and Isoprotic Point

The mean valence due to proton binding is defined as

$$z = \frac{C_+ - C_-}{C} = \frac{h^2 - K'_1 K'_2}{h^2 + hK'_1 + K'_1 K'_2} \tag{14}$$

This is zero at hydrogen ion activity h_i, satisfying the equation

$$h_i^2 = K'_1 K'_2 \tag{15}$$

Since $p = -\log$, one obtains the isoprotic pH:

$$(\text{pH})_i = (pK'_1 + pK'_2)/2 \tag{16}$$

Buffer Capacity of Isoprotic Ampholytes

The specific buffer capacity B of a weak protolyte is defined by the expression

$$B = (1/m)[dn/d(\text{pH})] \tag{17}$$

where m is the amount of weak protolyte and n the amount of alkali. For a monovalent weak acid:

$$B = \alpha(1 - \alpha)\ln 10 \tag{18}$$

where α is the degree of dissociation. A maximum is reached

$$B = (\ln 10)/4 \tag{19}$$

for $\alpha = 0.5$ corresponding to pH = pK' for a weak acid. The specific buffer capacity for a protolyte with an arbitrary number of protolytic groups can be shown to be identical with the derivative

$$B = -dz/d(\text{pH}) \tag{20}$$

where z is the mean valence (Eq. 14). Thus the buffer capacity becomes

$$B = \frac{1}{C} \frac{d(C_- - C_+)}{d(\text{pH})} \tag{21}$$

or

$$B = \frac{d}{d(\text{pH})} \left[\frac{K'_1 K'_2 - h^2}{h^2 + hK'_1 + K'_1 K'_2} \right] \tag{22}$$

After differentiation and insertion of the isoprotic condition (Eq. 15), we obtain

$$B_i = \ln 10/(1 + \sqrt{K'_1/4K'_2}) \tag{23}$$

Division of Eq. (23) by Eq. (19) gives the relative buffer capacity b_i in the isoprotic state, i.e., the capacity in units of the maximum capacity of a monovalent weak protolyte:

$$b_i = 4/(1 + \sqrt{K'_1/4K'_2}) \tag{24}$$

Because a bivalent protolyte cannot exert a buffer action better than twice that of a monovalent one, the maximum value of b_i must be 2. This leads to the conditions:

$$K'_1 \geqq 4K'_2 \quad \text{and} \quad \Delta pK' \geqq \log 4 \tag{25}$$

Conductivity of Isoprotic Ampholytes

The degree of ionization of an isoprotic ampholyte can be defined as

$$\alpha = \frac{C_+ + C_-}{C} = \frac{h^2 + K'_1 K'_2}{h^2 + hK'_1 + K'_1 K'_2} \tag{26}$$

It is unity at very low as well as at very high pH values. It also has a minimum at the isoprotic point which can be found by insertion of Eq. (15) into Eq. (26):

$$\alpha_i = 1/(1 + \sqrt{K'_1/4K'_2}) \tag{27}$$

Comparison of Eq. (24) and (27) produces

$$b_i = 4\alpha_i \tag{28}$$

which shows that a high degree of ionization (good conductivity) is accompanied by good buffering capacity and vice versa.

Carrier Ampholytes

The buffer capacity of carrier ampholytes at the isoprotic state (Eq. 24) is very important because they should exhibit a buffer action stronger than that of the proteins and therefore dictate the pH gradient. In addition it is necessary that the carrier ampholytes have appreciable conductivity in the isoelectric state and in the pH region around the neutral point in order to avoid local overheating and absorbance of the applied voltage in areas of low conductivity. This condition (i.e., local low conductivity) may reduce the field strength in conducting zones and in excess may abolish electrolytic transport and therefore focusing. Since the conductivity contribution of an isoprotic ampholyte is proportional to α_i (degree of ionization, Eq. 27), it is evident that ampholytes with a low pK' difference have adequate conductivity in the isoprotic state. Thus histidine ($\Delta pK' = 3.0$), glutamic acid ($\Delta pK' = 2.1$), and lysine ($\Delta pK' = 1.6$) can be considered useful carrier ampholytes whereas glycine ($\Delta pK' = 7.4$, corresponding to a degree of ionization of 0.00038 in the isoprotic state) is useless.

In order to be useful for isoelectric focusing experiments, the carrier ampholytes should contain a large number of isoelectric species differing less than 0.1 pH unit and should preferably be able to cover the range of pH 2.5 to 11.0. Such species should produce overlapping isoelectric distributions and therefore an approximately linear and smooth pH gradient. Other desirable properties of the ampholytes include good solubility in water, absence of hydrophobic groups, and low UV absorbance, especially at 280 nm. Vesterberg's (19) synthetic procedure was designed to produce a large number of isomers and homologs by coupling residues containing a carboxylic group to suitable amines. The resulting polyaminopolycarboxylic acids fulfill satisfactorily some of the requirements as expressed above and have found wide use in the isoelectric separation of proteins. However, these carrier ampholytes are far from ideal because (a) they produce a nonuniform conductance course, (b) their concentration differs throughout the pH gradient, (c) the pH gradient is not strictly linear, and (d) they may bind to certain proteins. These synthetic ampholytes are commercially available from LKB and cover the following pH ranges: 3.5–10.0, 2.5–4.0, 3.5–5.0, 4.0–6.0, 5.0–7.0, 5.0–8.0, 6.0–8.0, 7.0–9.0, 8.0–9.5, and 9.0–11.0.

The Steady State

Svensson (17) described the concentration distribution of an electrolyte at the isoelectric point as an "equilibrium" between mass transport and diffusional flow.

$$CME = D(dC/dx) \tag{29}$$

where C is the protein concentration, M is its mobility, E is the field strength, D is its diffusion coefficient, and x is the coordinate along the direction of current. The mobility M can be regarded as a linear function of x because of the narrowness of the zone at the pI. With the introduction of the proportionality factor p such that

$$M = -px \tag{30}$$

Eq. (29) can be written

$$dC/C = (Ep/D)xdx \tag{31}$$

If E, p, and D are treated as constants, this equation is integrated to give

$$C = C(0) \exp(-pEx^2/2D) \tag{32}$$

which expresses a Gaussian concentration distribution with a standard deviation (σ):

$$\sigma = (D/pE)^{1/2} \tag{33}$$

The proportionality factor p may be written as a derivative:

$$p = dM/dx = [-dM/d(\mathrm{pH})][d(\mathrm{pH})/dx] \tag{34}$$

The Transient State

Continuous analytical scanning isoelectric focusing techniques (43) have made possible the accurate estimation of the first and second moments of the concentration profile repetitively throughout the course of the isoelectric focusing experiment. This suggested the possibility of measuring the parameters D, p, and $dM/d(\mathrm{pH})$ through a mathematical analysis of the kinetics of isoelectric focusing. The method of TRANSIF (49) was thus born, promising to provide considerably more information than the analysis of the steady-state distribution alone. Weiss, Catsimpoolas, and Rodbard (40) presented a restricted theory of the kinetics of the new method. The principal assumption is that the mobility of the protein is a linear function of position at all times. The TRANSIF method is assumed to consist of three stages:

(1) Focusing, in which the system is allowed to approach the steady-state distribution for a time t_1.

(2)   Defocusing, in which the electrical field is abolished for a time t_2. This is assumed to be a pure diffusion process.

(3)   Refocusing for a time t_3, in which the field is reapplied and the distribution again approaches the steady-state.

The following assumptions have been adopted to permit a first theoretical approximation:

1.   A linear pH gradient is established prior to application of the sample protein. Alternatively, we may assume that the pH gradient is formed very rapidly compared with the kinetics of focusing of the macromolecule of interest.

2.   The pH mobility curve of the protein is assumed to be linear. This assumption is valid only for a limited region near the isoelectric point. On the basis of these two assumptions, $p = dM/dx$ is constant. This single assumption could be used in lieu of the above.

3.   The electrical field strength (E) is assumed to be uniform throughout the entire separation path. In lieu of assumptions 1–3, we could simply assume that $pE = dv/dx$ is constant, where v is velocity.

4.   Diffusion and mobility coefficients are assumed to be independent of concentration.

5.   Diffusion coefficients are assumed to be independent of pH (at least in the region near the isoelectric point).

6.   It is assumed that there are no physical/chemical interactions between the protein and other chemical species present (e.g., ampholytes), and no self-association or protein–protein interactions).

7.   Band spreading is governed only by diffusion or by a diffusion–like process. Thus electrostatic effects are ignored and it is assumed that the protein is perfectly homogeneous with respect to pI, charge, mobility, radius, and diffusion coefficient.

8.   No perturbing phenomena such as electroendosmosis, convective disturbances, or precipitation at the isoelectric point are present.

9.   If a gel or density gradient is used as a supporting medium, their effects on diffusion coefficients and on mobility are neglibible (or at least constant throughout the gel), and there is no effect on the uniformity of the electrical field. Thus the effect of the viscosity gradient which is superimposed on the density gradient in sucrose–gradient columns is ignored. Likewise, the molecular sieving effects which are present when polyacrylamide gels are used as a supportive medium are ignored.

10.   The effect of the boundary condition that there can be no flux

of the species of interest through the ends of the gel column, or that there is an abrupt discontinuity of pH at the ends of the column, is ignored. These effects should become insignificant shortly after the start of the experiment.

The above assumptions may be relaxed later to provide a more generalized and practical theory. For experimental purposes it is sufficient and convenient to find $\mu_1(\tau)$ and $\sigma^2(\tau)$, i.e., the mean and square of the standard deviation of peak width. These equations are:

1. Focusing: $0 \leqq \tau \leqq \tau_1$

$$\mu_1(\tau) = \mu_1(0)e^{-\tau} + y_0(1 - e^{-\tau}) \tag{35}$$

$$\sigma^2(\tau) = \sigma^2(0)e^{-2\tau} + \alpha(1 - e^{-2\tau}) \tag{36}$$

2. Defocusing: $\tau_1 \leqq \tau \leqq \tau_1 + \tau_2$

$$\mu_1(\tau) = \mu(0)e^{-\tau_1} + y_0(1 - e^{-\tau_1}) = \text{constant} \tag{37}$$

$$\sigma^2(\tau) = \sigma^2(0)e^{-2\tau_1} + \alpha(1 - e^{-2\tau_1}) + 2\alpha(\tau - \tau_1) \tag{38}$$

3. Refocusing: $\tau_1 + \tau_2 \leqq \tau$

$$\mu_1(\tau) = \{\mu_1(0)e^{-\tau_1} + y_0(1 - e^{-\tau_1})\} \exp \{-(\tau - \tau_1 - \tau_2)\}$$
$$+ y_0(1 - \exp \{-(\tau - \tau_1 - \tau_2)\}) \tag{39}$$

$$\sigma^2(\tau) = \{\sigma^2(0)e^{-2\tau_1} + \alpha(1 - e^{-2\tau_1}) + 2\alpha\tau_2\}$$
$$\times \exp \{-2(\tau - \tau_1 - \tau_2)\}$$
$$+ \alpha(1 - \exp \{-2(\tau - \tau_1 - \tau_2)\}) \tag{40}$$

where L is the column length
x_0 is the position of the isoelectric point
$\alpha = D/(L^2 pE)$
$\tau = pEt$
$y_0 = x_0/L$

A computer simulation study derived from theory of the time course of the centroid (μ) and σ^2 in TRANSIF is shown in Fig. 1.

The centroid approaches the isoelectric point by an exponential decay during focusing and refocusing. With ideal initial pulse loading, the bandwidth (σ^2) increases during focusing, asymptotically approaching the steady–state value.

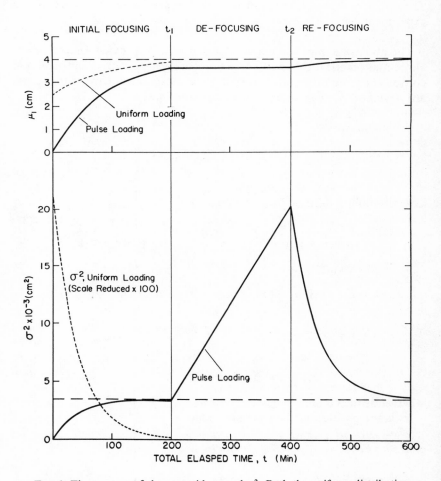

FIG. 1. Time course of the centroid, μ, and σ^2. Both the uniform distribution and "pulse loading" cases are shown for the initial focusing stage.

Parameters: $L=5$ cm; $X_0=4$ cm; $D=7\times10^{-7}$ cm^2/sec; $pE=20\times10^{-5}=2\times10^{-4}$ sec^{-1}; $E=20$ V/cm; $P=10^{-5}$ cm/(sec) (V).

ELECTROOPTICAL SCANNING

Electrooptical Scanning in Isoelectric Focusing: Advantages and Limitations

Biomolecules separated by isoelectric focusing have been detected and analyzed by the following methods: (a) optical and electrooptical *in situ* methods (e.g., visible and UV light absorption); (b) nonspecific dye staining methods (e.g., Coomassie Blue staining of proteins); (c) specific dye staining methods (e.g., for glycoproteins and lipoproteins); (d) immunochemical methods (utilizing antigen–antibody reactions); (e) enzymatic activity methods; (f) specific biological activity methods (bioassay); (g) radioactivity labeling methods; and (h) fluorescent labeling methods.

From all the above techniques, only optical and especially electrooptical methods allow for the continuous monitoring of a particular separation as a function of time without interruption of the electric field. In addition, electrooptical scanning methods lend themselves to precise mathematical peak analysis for the evaluation of methodological parameters and the measurement of physical constants. In particular, reference to the isoelectric focusing and isotachophoresis techniques which require the attainment of a "steady-state," continuous monitoring of the separation field is especially advantageous for evaluation of the minimal focusing time (t_{MF}) in isoelectric focusing of the "constancy of zone length" in isotachophoresis.

To date, the main limitations of the electrooptical detection method are the nonspecificity of detection and reduced sensitivity in comparison to some other detection techniques (e.g., immunological and enzymatic). However, proteins in amounts ranging from 10 to 100 μg can be quantitatively detected by the UV scanning method (at 280 nm). The sensitivity can be increased tenfold if the absorbance at 220 nm is recorded, but this precludes the use of a polyacrylamide gel as a supporting medium. The absorbance at 220 nm is largely due to the peptide bonds and is relatively little affected by the content of aromatic amino acids such as tryptophan and phenylalanine. Therefore, this particular wavelength could be preferentially used for separation in sucrose density gradient. The sensitivity of UV detection at 280 nm can be improved—within 1 to 2 orders of magnitude—by an apparatus which will be utilizing a double-wavelength ratio-recording system to significantly reduce baseline noise and decrease the lower limits of detection.

Principle of in Situ Visible or UV Scanning

Electrophoretic separation is performed in a quartz cell which moves perpendicularly to a thin (25 to 50 μ slit-width) light beam. When the beam encounters separated zones of the sample, it is absorbed proportionally to the amount of material present in the zone and its extinction coefficient at a particular wavelength. A photomultiplier placed behind the quartz column detects the light variation and produces a current which can be converted electronically to an analog voltage related linearly to the zone profile absorbance. The analog voltage can be either recorded directly on a strip-chart recorder or preferentially can be digitized and processed by a computer. The separation path is scanned continuously during the electrophoretic run, resulting in several electropherograms (scans) as a function of time. The sample can be scanned at any preferred wavelength in the 200 to 800 nm range, or the scanner can be stopped and one particular zone can be scanned as a function of wavelength. The electric field is applied at all times. However, the current can be interrupted and the diffusion of the zone can be followed by the broadening of the concentration distribution with time. Thus both electrophoretic and diffusional mass transport phenomena can be evaluated. This is a very important feature in the digital measurement of physical constants of biomolecules and methodological parameters as will be discussed below.

Coupling of the scanning instrument to a digital data acquisition and processing system allows direct measurement of the zeroeth (m_0), reduced first (m'_1), and second (m_2) statistical moments of the zone concentration profile which correspond, respectively, to the area, the position ($\bar{x}$), and the variance (σ^2) of the peak. The experimental values of m_0, σ^2, and $\sigma \, (= \sqrt{\sigma^2})$ in conjunction with $\bar{x}$ values of pI markers are the only parameters required to be measured in order to obtain valuable quantitative information in TRANSIF experiments. These techniques have been described in detail by Catsimpoolas and co-workers (*35–43*).

METHODOLOGICAL PARAMETERS

Determination of Minimal Focusing Time

In the simple case of a single protein uniformly distributed in the column before the electric field is applied, the minimal time required to obtain complete focusing can be determined by following the position ($\bar{x}$) of the discernible peaks migrating from the two ends of the path (positive and negative) toward the isoelectric point position where they merge into one peak. At the steady-state and in the absence of significant pH gradient instability, the peak position in pI should remain constant with time. Ideally, the peak area (zeroeth moment) and the variance (second moment about the mean) should also remain constant when the steady-state is reached. The latter two parameters can be employed in evaluating "steady-state" conditions (and therefore minimal focusing time) of a mixture of proteins. It should be emphasized that depending on their individual pH–mobility relationship, proteins in a mixture may approach the "steady-state" at different times. Other factors that affect the minimal focusing time of individual proteins are: (a) sieving effects in gels, (b) nonuniform electric field strength, and (c) presence of a viscosity gradient (e.g., sucrose density gradient). The minimal focusing time may be also generally affected by: (a) ampholyte concentration, (b) electric field strength, (c) pH range of carrier ampholytes, (d) temperature, and (e) presence of additives, e.g., urea.

Segmental pH Gradient

This parameter, $\Delta(\text{pH})/\Delta x$ (cm^{-1}), is measured using two pI markers of closely spaced isoelectric points from

$$\frac{\Delta(\text{pH})}{\Delta x} = \frac{\text{pI}_\text{A} - \text{pI}_\text{B}}{\bar{x}_\text{A} - \bar{x}_\text{B}} \tag{41}$$

where pI is the isoelectric point, $\bar{x}$ is the peak position, and subscripts A and B denote two pI markers. In using the above equation, it is assumed that species A and B have reached their isoelectric point, and that $\Delta(\text{pH})/\Delta x$ is constant between pI_A and pI_B where $\bar{x}_\text{A} - \bar{x}_\text{B}$ represents a small segment of the separation path.

Resolution

Arbitrarily assigning resolution of unity to a just resolved double-zone, the resolution R_s can be expressed as

$$R_s = \frac{\Delta \bar{x}}{1.5(\sigma_\text{A} + \sigma_\text{B})} \tag{42}$$

where $\Delta \bar{x}$ is the peak separation of two zones A and B with standard deviations of σ_A and σ_B. Again, $\Delta \bar{x}, \sigma_\text{A}$, and σ_B can be measured directly at any stage of fractionation.

Resolving Power

In isoelectric focusing the resolving power has been defined by Vesterberg and Svensson to be

$$\Delta pI = 3[d(pH)/dx]\sigma \tag{43}$$

Since $\Delta(pH)/\Delta x$ and σ can be obtained digitally in TRANSIF, the resolving power can be estimated directly.

APPARENT PHYSICAL CONSTANTS

Apparent Isoelectric Point

If an "unknown" protein (U) is included in the segmental pH gradient as described above, its apparent isoelectric point can be calculated by

$$pI_U = pI_A + \left(\frac{\Delta(pH)}{\Delta x}\right)(\bar{x}_A - \bar{x}_U) \tag{44}$$

All three species A, B, and U should be at pH equilibrium, i.e., at the steady-state.

Diffusion in Polyacrylamide Gels: Determination of the Retardation Coefficient (C_R)

The apparent diffusion coefficient in polyacrylamide gels is related to the gel concentration by

$$\log D = \log(D_0) - C_R T \tag{45}$$

where D_0 is the free diffusion coefficient, T is the gel concentration, D the apparent diffusion coefficient at any gel concentration T, and C_R is the retardation coefficient obtained from diffusion data. C_R can be measured during the defocusing stage of TRANSIF experiments from the slope of the plot log D vs T. Thus TRANSIF in polyacrylamide gels can provide a measure of molecular size. It is therefore possible that the effective molecular radius $\bar{R}$ and MW could be estimated by the present method from plots of C_R vs $\bar{R}$, or vs MW in analogy to the Rodbard-Chrambach plots (50).

Measurement of D and pE

As mentioned above, a TRANSIF experiment is characterized by three stages; namely, focusing, defocusing, and refocusing. In the focusing stage the sample is subjected to electrophoresis on a pH gradient for time t_1 until a nearly steady-state distribution of the focused zone is achieved. In the defocusing stage, the electrical field is removed for time t_2, allowing the zone to spread by diffusion. In the refocusing stage, the electrical field is reapplied for time t_3 and the distribution reapproaches the steady-state. The advantages of performing kinetic analysis of zone focusing during the refocusing period are: (a) focusing is carried out by starting with a (near) Gaussian distribution; (b) the zone is restricted to a narrow region of the pH (and mobility) spectrum near the isoelectric point; and (c) data are collected under conditions of nearly linear pH gradient $(d(\text{pH})/dx)$ and linear pH–mobility curve $[dM/d(\text{pH})]$. Thus, if the parameters $d(\text{pH})/dx$ and $dM/d(\text{pH})$ are constant, the experimentally measurable parameter p will also remain constant throughout the refocusing experiment, since

$$p = [dM/d(\text{pH})][d(\text{pH}/dx] \tag{46}$$

The parameter i is related to the standard deviation of the concentration distribution of a focused zone at the steady-state by

$$\sigma = \sqrt{D/pE} \tag{47}$$

where

$$E = i/qK \tag{48}$$

where i is the current, q is the cross-sectional area, k is the conductance, and E is electric field strength.

For experimental purposes, the kinetics of defocusing and refocusing can be evaluated by following the changes of σ^2, which is the square of the standard deviation of peak width, vs elapsed time. The equations describing the behavior of σ^2 during these two stages of the experiment have been derived from theory (2) to be:

1. *Defocusing*

$$\sigma^2(t_2) = \sigma^2(t_1) + 2D(t_2 - t_1) \tag{49}$$

2. *Refocusing*

$$\sigma^2(t) = (D/pE) + 2Dt_2 \exp(-2pEt_3) \tag{50}$$

Experimentally, a plot of σ^2 vs $2t_2$ should permit estimation of the apparent diffusion coefficient D (as the slope of the line) during the defocusing stage. Also a plot of $\log[(\sigma_R^2 - \sigma_F^2)/\sigma_D^2]$ vs $2t_3$ during the refocusing stage can be used to determine the parameter pE as the slope of the linear plot. If $d(\text{pH})/dx$ and E are known, the physical constant $dM/d(\text{pH})$ can be estimated from Eq. (46).

CONCLUDING REMARKS

Although isoelectric focusing has been successfully applied to the separation of amphoteric molecules (51), especially proteins, as evidenced by more than 1500 published articles to-date, the full potential of the technique has not yet been exploited. This is primarily due to the lack of suitable carrier ampholytes, which should provide a uniform conductance and concentration distribution course throughout the separation path, and a stable linear pH gradient. The commercially available carrier ampholytes fall short in all three of the above desirable properties. The success of steady-state isoelectric focusing—as commonly practiced today—lies primarily in the relative reproducibility of the isoelectric point and the ability to separate different isoelectric species with very high resolution. However, with the development of the TRANSIF technique, it should be possible to utilize the method not only for the separation but also for the physicochemical characterization of amphoteric molecules. To be specific, one should be able to use TRANSIF in obtaining the diffusion coefficient, the $dM/d(\text{pH})$ coefficient, the isoelectric point, and the pH–mobility curve of a protein with a good degree of confidence in the results. The kinetic theory and available instrumentation allow us to do the necessary measurements, but the compiling of correction factors due to nonideal effects, stemming primarily from the present imperfection of carrier ampholytes, renders the method impractical. Some of the corrections that have to be made involve the effect of ampholyte concentration and zone conductance on the measurable physical constants. These should have been relatively easy to carry out if there was assurance that these corrections apply uniformly throughout the column. Other corrections involve "zone load," viscosity, and temperature effects.

Despite the present shortcomings, isoelectric focusing has become an established separation technique with a very promising future both at the preparative and analytical levels. Dynamic development of new methodology and instrumentation coupled with a selective extension of the kinetic theory and the much needed synthesis of "second generation" ampholytes will undoubtedly suggest new avenues of application.

REFERENCES

1. H. Picton and S. E. Linder, *J. Chem. Soc. 71*, 568 (1897).
2. K. Ikeda and S. Suzuki, U. S. Patent 1,015,891 (1912).
3. R. R. Williams and R. E. Waterman, *Proc. Soc. Exp. Biol. Med.*, 27, 56 (1929).

4. R. J. Williams, *J. Biol. Chem.*, *110*, 589 (1935).
5. V. du Vigneaud, G. W. Irwing, H. M. Dyer, and R. R. Sealock, *Ibid.*, *123*, 45 (1938).
6. A. Tiselius, *Svensk Kem. Tidskr.*, *53*, 305 (1941).
7. R. L. H. Synge, *Biochem. J.*, *49*, 642 (1951).
8. F. L. Sanger and H. Tuppy, *Ibid.*, *49*, 463 (1951).
9. A. Kolin, *J. Chem. Phys.*, *22*, 1628 (1954).
10. A. Kolin, *Ibid.*, *23*, 417 (1955).
11. A. Kolin, *Proc. Nat. Acad. Sci.*, *U.S.*, *41*, 101 (1955).
12. A. Kolin, *Meth. Biochem. Anal.*, *6*, 259 (1958).
13. H. Hoch and G. H. Barr, *Science*, *122*, 243 (1955).
14. H. J. MacDonald and M. B. Williamson, *Naturwissenschaften*, *42*, 461 (1955).
15. J. R. Maher, W. O. Trendle, and R. L. Schultz, *Ibid.*, *43*, 423 (1956).
16. A. H. Tuttle, *J. Lab. Clin. Med.*, *47*, 811 (1956).
17. H. Svensson, *Acta Chem. Scand.*, *15*, 425 (1961).
18. H. Svensson, *Ibid.*, *16*, 456 (1962).
19. O. Vesterberg, *Ibid.*, *23*, 2653 (1969).
20. O. Vesterberg and H. Svensson, *Ibid.*, *20*, 820 (1966).
21. Z. L. Awdeh, A. R. Williamson, and B. A. Askonas, *Nature*, *219*, 66 (1968).
22. N. Catsimpoolas, *Anal. Biochem.*, *26*, 480 (1968).
23. G. Dale and A. L. Latner, *Lancet*, *1*, 847 (1968).
24. J. S. Fawcett, *FEBS Lett.*, *1*, 81 (1968).
25. R. F. Riley and M. K. Coleman, *J. Lab. Clin. Med.*, *72*, 714 (1968).
26. C. W. Wrigley, *J. Chromatogr.*, *36*, 362 (1968).
27. D. H. Leaback and A. C. Rutter, *Biochem. Biophys. Res. Commun.*, *32*, 447 (1968).
28. M. B. Hayes and D. Wellner, *J. Biol. Chem.*, *244*, 6636 (1969).
29. N. Catsimpoolas, *Immunochemistry*, *6*, 501 (1969).
30. N. Catsimpoolas, *Biochim. Biophys. Acta*, *175*, 214 (1969).
31. N. Catsimpoolas, *Clin. Chim. Acta*, *23*, 237 (1969).
32. N. Catsimpoolas, *Science Tools*, *16*, 1 (1969).
33. N. Catsimpoolas, *Separ. Sci.*, *5*, 523 (1970).
34. N. Catsimpoolas, *Ibid.*, *8*, 71 (1973).
35. N. Catsimpoolas, *Anal. Biochem.*, *54*, 66 (1973).
36. N. Catsimpoolas, *Ibid.*, *54*, 88 (1973).
37. N. Catsimpoolas, *Ibid.*, *54*, 79 (1973).
38. N. Catsimpoolas, in *Isoelectric Focusing* (J. P. Arbuthnott, ed.), Butterworths, London, In Press.
39. N. Catsimpoolas and A. L. Griffith, *Anal. Biochem.*, *56*, 100 (1973).
40. G. H. Weiss, N. Catsimpoolas, and D. Rodbard, *Arch. Biochem. Biophys.*, *163*, 106 (1974).
41. N. Catsimpoolas, W. W. Yotis, A. L. Griffith, and D. Rodbard, *Ibid.*, *163*, 113 (1974).
42. N. Catsimpoolas, B. E. Campbell, and A. L. Griffith, *Biochim. Biophys. Acta*, *351*, 196 (1974).
43. N. Catsimpoolas, *Ann. N. Y. Acad. Sci.*, *209*, 65 (1973).
44. R. A. Alberty, in *The Proteins*, Vol. 1 (A. H. Neurath and K. Bailey, eds.), Academic, New York, 1953, p. 461.
45. S. P. L. Sørensen, K. Linderstrøm-Lang, and E. Lund, *Compt. Rend. Trav. Lab. Carlsberg*, *16*, 5 (1926).

46. R. K. Cannan, *Chem. Rev.*, *30*, 295 (1942).
47. D. Davidson, *J. Chem. Educ.*, *32*, 550 (1955).
48. H. Rilbe, *Ann. N. Y. Acad. Sci.*, *209*, 11 (1973).
49. N. Catsimpoolas, *Fed. Proc.*, *32*, 625 (1973).
50. D. Rodbard and A. Chrambach, *Anal. Biochem.*, *40*, 95 (1971).
51. N. Catsimpoolas, "Isoelectric Focusing and Isotachophoresis," *Ann. N. Y. Acad. Sci.*, *209*, 1–529 (1973).

Foam Separation of Anions from Aqueous Solution: Selectivity of Cationic Surfactants

R. B. GRIEVES, W. CHAREWICZ,* and P. J. W. THE

THE UNIVERSITY OF KENTUCKY
LEXINGTON, KENTUCKY 40506

Abstract

Anions are selectively separated and concentrated from dilute aqueous solution by foam fractionation. Selectivity coefficients are established from steady-state equilibrium data (solution concentrations 10^{-4} to 10^{-3} M) for SCN^-, I^-, ClO_3^-, NO_3^-, BrO_3^-, and NO_2^-, each vs Br^-, with the quaternary ammonium surfactant modeled as a soluble anion exchanger. Studies are reviewed on the foam separation of Re(VII), Mo(VI), and V(V) oxyanions; Au(I), Ag(I), Ni(II), and Co(III) cyanide complexes; and Pt(IV), Pd(II), and Au(III) chloro complexes. In a five-component system, the oxyanions of Re(VII), Mo(VI), Cr(VI), W(VI), and V(V) are foam fractionated from 10^{-6} M solutions with the cationic surfactant, hexadecyldimethylbenzylammonium chloride. In the batch, time-dependent experiments, the metals are monitored by radiotracers and gamma radiation spectrometry. At pH 6.0 and a chloride (NaCl) concentration of 10^{-2} M, and at pH 2.0, adjusted with HCl, Re(VII) and Mo(VI) oxyanions can be separated completely from Cr(VI), W(VI), and V(V) oxyanions. The selectivity sequences are discussed in terms of acid-base equilibria and in terms of the absolute partial molal entropy of each anion in aqueous solution, as a measure of surfactant cation–anion interaction.

INTRODUCTION

Foam separation processes have been used to separate and concentrate inorganic anions from dilute aqueous solution; a cationic surfactant is utilized which interacts preferentially with one or more anions (colligends)

*Institute of Inorganic Chemistry and Metallurgy of Rare Elements, Technical University of Wroclaw, 50-370, Wroclaw, Poland.

compared to competing anions. The effectiveness of the process is determined by the selectivity of the surfactant toward the colligend(s). The surfactant–colligend interaction may occur in the bulk solution and/or at the air–solution interfaces of generated gas bubbles. The surfactant–colligend ion pair or soluble complex is concentrated in the foam which is produced atop the bulk solution. In the absence of particles formed among the surfactant–colligend pairs, the process is termed foam fractionation (1).

The principles of foam fractionation have been detailed in several reviews (1–12) which have included references to a great number of specific separations, generally from dilute aqueous solutions (and involving the discussion of a considerable amount of solution chemistry). In these studies, effects have been enumerated of solution variables such as concentrations of surfactant and colligend, pH, ionic strength, and the presence of specific ionic species which may modify the charge and structure of the colligend or compete with the colligend for the surfactant; of operational variables such as aeration rate, gas bubble size, foaming time, and foam column geometry; and of the mode of operation: batch or continuous flow, with various schemes of foam recycle, feed position modification, etc.

Recently, a cationic surfactant concentrated from dilute aqueous solution at air–solution bubble interfaces has been modeled as a soluble ion exchanger, and selectivity coefficients have been established for SCN^-, I^-, ClO_3^-, NO_3^-, BrO_3^-, and NO_2^-, each vs Br^- (13, 14). Several investigations have been conducted to determine the selectivity of cationic surfactants for transition metal oxyanions (15–21), leading to the successful separation and fractionation of as many as five transition metals from a dilute (1.0×10^{-6} M) solution of their oxygen compounds (22, 23). Efforts have been made at selectivity prediction, based on the thermodynamic properties of the anions (14, 19, 20, 24).

SELECTIVITY COEFFICIENTS OF EACH OF A SERIES OF ANIONS VS BROMIDE FROM CONTINUOUS FLOW STUDIES

Steady-state, single equilibrium stage experiments were conducted in the foam fractionation unit shown schematically in Fig. 1 (13, 14). For a feed stream containing, for example, NaI of concentration c_i in I^- and the quaternary ammonium surfactant, ethylhexadecyldimethylammonium bromide (EHDA–Br), of concentration e_i in $EHDA^+$ and b_i in Br^- (NaBr also was present in some experiments), there were produced by aeration of the bulk solution a steady-state residual stream lean in surfactant and a steady-state foam stream rich in surfactant. The residual stream concentrations are designated c_r in I^-, e_r in $EHDA^+$, and b_r in Br^-. The as-

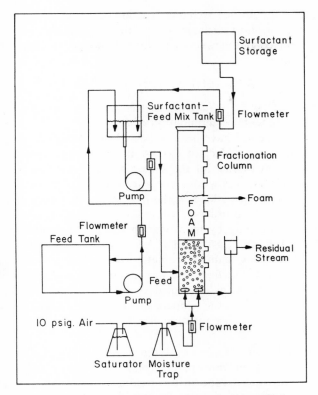

FIG. 1. Schematic diagram of continuous flow, single equilibrium stage foam separation unit.

sumption could be made (*13*) that the foam consisted of entrained bulk liquid (of concentration c_r, e_r, and b_r) in equilibrium with surface liquid containing surfactant of surface concentration Γ_e, plus the fixed and diffuse layers of counterions, of surface concentrations $\Gamma_c(I^-)$ and $\Gamma_b(Br^-)$.

As each bubble rose through the bulk solution, there may have occurred an exchange reaction

$$(EHDA–Br)_s + (I^-)_r \rightleftarrows (EHDA–I)_s + (Br^-)_r$$

in which the subscript s designates the surface layer or phase. The selectivity coefficient is defined by

$$K' = \Gamma_c b_r / \Gamma_b c_r \tag{1}$$

and it was shown that the liquid height in the column had no effect on

K' (*13*). Either the surface exchange reaction occurred rapidly and reached equilibrium at a short distance above the air diffusers, or no surface exchange occurred and the selectivity was determined by ion pair formation in the bulk solution:

$$K' = K'' = (EHDA–I)_r(Br^-)_r/(EHDA–Br)_r(I^-)_r \qquad (2)$$

The ratio of Γ_c/Γ_b in Eq. (1) can be replaced by $(c_i - c_r)/(b_i - b_r)$ to enable the direct experimental determination of K'.

Experimental details have been presented previously (*13, 14*). The feed concentration of the surfactant EHDA–Br, e_i, ranged from 1.0 to 4.0 × 10^{-4} M, and that of each of the salts NaSCN, NaI, NaClO$_3$, NaNO$_3$, NaBrO$_3$, or NaNO$_2$ was within the range 0.4 to 6.6 × 10^{-4} M. In some experiments, NaBr was added to the feed stream in concentrations from 1.0 to 7.8 × 10^{-4} M in order to reduce the c_r/b_r ratio to values lower than could be achieved only by maneuvering the feed concentrations of the colligend salt and of EHDA–Br. The feed stream was within the pH range 5.6 to 5.8. The Pyrex foam fractionation column was 89.5 cm high and 9.7 cm in diameter. The feed rate was maintained at 0.056 liter/min, the air rate through the twin sintered glass diffusers of 50 μm porosity was held at 0.4 liter/min, and the temperature was 24 ± 1.0°C. The surfactant was analyzed by two-phase titration and the various anions by UV absorption (*14*).

Experimental results for each of the anionic colligends are presented in Fig. 2, in which the ratio of the colligend to bromide in the surface phase is related to that ratio in the residual stream. All data points are presented for SCN$^-$, I$^-$, and NO$_2$$^-$, while for ClO$_3$$^-$, NO$_3$$^-$, and BrO$_3$$^-$, due to crowding, points are given only at e_i = 1.8 × 10^{-4} M. For ClO$_3$$^-$ and BrO$_3$$^-$, data were also taken at e_i = 1.0 × 10^{-4} M and 4.0 × 10^{-4} M and for NO$_3$$^-$ at 1.4, 1.6, and 2.0 × 10^{-4} M. All data were used to establish the straight lines, and K' was independent of e_i. The resultant values of K', together with statistical indicators of the goodness of fit, are given in Table 1. The last column in Table 1 is discussed in a later section.

The values of Γ_c/Γ_b for SCN$^-$ at c_r/b_r > 0.35 and for I$^-$ at c_r/b_r > 1.0 (for e_i ≐ 1.8 × 10^{-4} M) began to rise sharply. For both of these systems, fine particles were observed in the foam at the foam removal port; K' would not be constant if EHDA–SCN or EHDA–I were precipitated in the surface phase (*14*). The effect on K' of the activity coefficient ratio in the concentrated surface layer or phase for each of the anionic colligends was determined by relating K' to Γ_c/Γ_e, the fraction of the exchanger occupied by the preferred ion. K' was found to be independent of Γ_c/Γ_e, indicating a constant activity coefficient ratio, and was found to be rela-

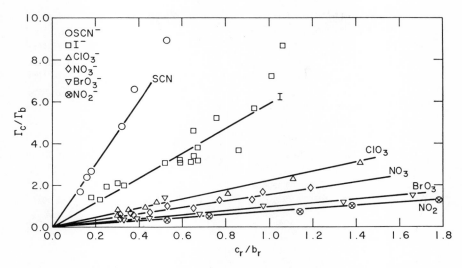

FIG. 2. Establishment of selectivity coefficients for SCN^-, I^-, ClO_3^-, NO_3^-, BrO_3^-, and NO_2^- vs Br^-.

TABLE 1

Determination of Selectivity Coefficients of Anionic Colligends vs Bromide

Anionic colligend	K'	Total number of data points	95% Confidence limits for K'	Correlation coefficient, r, for K'	$\bar{S}_{298}^2$ abs (eu)	
SCN^-	15.1	4	15.1 (1.0 ± 0.29)	0.98	41.5	
I^-	5.85	17	5.85 (1.0 ± 0.31)	0.87	31.6	
ClO_3^-	2.21	25	2.21 (1.0 ± 0.058)	0.98		44.5
NO_3^-	1.56	23	1.56 (1.0 ± 0.18)	0.93		40.5
Br^-	1.00				24.8	
BrO_3^-	0.95	21	0.95 (1.0 ± 0.084)	0.98		44.0
NO_2^-	0.73	5	0.73 (1.0 ± 0.33)	0.98		35.4

tively independent of ionic strength (*14*). Based on a limited amount of data, K' for NO_3^-/Br^- was found to increase with decreasing surfactant chain length for C_{14}, C_{16}, and C_{18} trimethylammonium bromides (*14*).

REVIEW OF FOAM SEPARATION STUDIES IN POLAND: ANIONIC COLLIGENDS AND CATIONIC SURFACTANTS

Among the known elements, some of the transition metals occupy a peculiar position due to their value, recently increasing demands, and the lack of sufficiently rich mineral resources. In Poland, particular emphasis has been placed on foam separation processes to separate and concentrate Re(VII), Mo(VI), and V(V) oxyanions; Au(I), Ag(I), Ni(II), and Co(III) cyanide complexes; and Pt(IV), Pd(II), and Au(III) chloro complexes. Table 2 presents a review of pertinent references on Re(VII), Mo(VI), and V(V) oxygen compounds: this work is of particular interest due to the separation study of the five-colligend-system detailed in the next section. Utilizing two different surfactants, ReO_4^- could be separated rather completely from $HMoO_4^-$ in 1.0×10^{-2} M H_2SO_4 solution (17, 19). V(V) oxygen compounds, in 1.0×10^{-6} M solution, could be effectively foam separated with both an anionic and a cationic surfactant, depending on the H_2SO_4 concentration (18).

Additional, specific investigations have been made of the foam separa-

TABLE 2

Studies of Foam Separation of Re(VII), Mo(VI), and V(V) Oxygen Compounds

Solution composition	Transition metal oxygen compound concentration (M)	Surfactant	Surfactant concentration (M)	Ref.
ReO_4^-	1.0×10^{-4}	Hexadecyldiethyl-ammonium hydrochloride (HDEAHCl)	5.0×10^{-4}	25
MoO_4^{2-}	1.0×10^{-4}	HDEAHCl	3.0×10^{-4}	26
Mo(VI) oxygen compounds	1.0×10^{-6} to 1.4×10^{-3}	HDEAHCl	5.0×10^{-5} to 4.0×10^{-4}	15
ReO_4^-	1.0×10^{-5} to 3.0×10^{-4}	HDEAHCl	5.0×10^{-5} to 2.0×10^{-4}	16
$ReO_4^- + MoO_4^{2-} +$ H_2SO_4	1.9×10^{-4}	Dodecyldimethyl-benzylammonium bromide (DDMBABr)	6.0×10^{-4}	17
V(V) oxygen compounds + H_2SO_4	1.0×10^{-6} to 1.0×10^{-4}	Hexadecyltrimethyl-ammonium bromide (HTMABr) (also anionic surfactants)	5.0×10^{-4} to 7.5×10^{-4}	18
$ReO_4^- + MoO_4^{2-}$	2.0×10^{-5}	DDMBABr	2.0×10^{-4}	19
$ReO_4 + HPO_4^{2-}$	5.0×10^{-5}	HDEAHCl	8.0×10^{-4}	19
Re(VII), Mo(VI), and P(V) oxygen compounds	1.0×10^{-5} to 1.0×10^{-4}	DDMBABr	8.0×10^{-4}	20
ReO_4^-, MoO_4^{2-}	5.0×10^{-5}	HDEAHCl	2.0×10^{-4}	21

tion of $AuCl_4^-$ (27, 28); $PtCl_6^{2-}$, $PtCl_4^{2-}$, and $Pt(CN)_4^{2-}$ (29, 30); $Au(CN)_2^-$ and $Ag(CN)_2^-$ (31); $PdCl_4^{-2}$ (32); $Co(CN)_6^{3-}$ (33); $Co(SCN)_4^{2-}$ and $Ni(SCN)_4^{2-}$ (34); and anionic complexes of Ni(II), Th(IV), Pa(V), U(VI), Am(III), Cm(III), Eu(III), Tm(III), and Yb(III) (35–37). A solvent sublation process has been applied to ReO_4^- and MoO_4^{2-} (38). These investigations, almost entirely conducted in batch, time-dependent-based foam separation units, have included experimental methodology, effects of process-controlling variables, solution chemistry aspects, the determination of selectivity, and direct applications of the process to hydrometallurgy.

FOAM FRACTIONATION OF A FIVE COLLIGEND SYSTEM: OXYANIONS OF Re(VII), Mo(VI), Cr(VI), W(VI), AND V(V)

From a dilute (10^{-6} M) aqueous solution of the oxygen compounds of Re(VII), Mo(VI), Cr(VI), W(VI), and V(V), the foam separation selectivity should be established by two factors: first, acid-base equilibria in solution which can cause the increase in charge, the neutralization, and even the reversal of charge of the oxyanion, together with the modification of its structure. For these transition metal oxygen compounds, generally in acidic or neutral solution, the following equilibria can be written,

$$H_n MeO_m \overset{K_1}{\rightleftarrows} H^+ + H_{n-1} MeO_m^-$$

$$H_{n-1} MeO_m \overset{K_2}{\rightleftarrows} H^+ + H_{n-2} MeO_m^{2-}$$

for Mo(VI), Cr(VI), and W(VI), $n = 2$ and $m = 4$; for Re(VII), $n = 1$ and $m = 4$; and for V(V), $n = 3$ or 1 and $m = 4$ or 3.

Based upon literature references, selected values of pK_1 and pK_2 are presented in Table 3. In the 10^{-6} M solutions involved, complex anions and polynuclear species are neglected except for anionic chloro complexes of Mo(VI) which have been extensively reported (46, 57, 58). For V(V) the cation VO_2^+ has been reported in dilute acid solutions (54, 59, 60).

The second factor determining the selectivity should be specific, metal oxyanion–surfactant cation interaction (ion pair formation) in solution. For the system described, ReO_4^-, for example, must compete for the surfactant cations with the other four transition metal oxyanions and with other anions, such as Cl^- or SO_4^{2-}, which may be present in concentrations several orders of magnitude greater than that of ReO_4^-. The relative competitiveness of the various anionic species is determined by their charges, degrees of hydration, and structures.

TABLE 3

Selected Acid-Base Dissociation Constants for Transition Metal Oxyacids

Metal	Constants	Refs.
Re(VII)	$pK_1 = -1.2$	*39–41*
Mo(VI)	$pK_1 = 1.4$	*39, 42, 43*
	$pK_2 = 4.0$	*39, 42, 44, 45*
Cr(VI)	$pK_1 = -0.9$	*39, 41, 47–49*
	$pK_2 = 6.5$	*39, 41, 48, 49*
W(VI)	$pK_1 = 2.2–3.5$	*39, 50, 51*
	$pK_2 = 3.8–4.6$	*50–52*
V(V)	$pK_1 = 3.6$	*39, 53–55*
	$pK_2 = 8.2$	*39, 41, 53, 55, 56*
	$pK_3 = 13.5$	*41, 55, 56*

Experimental

A Pyrex batch flotation cell, 45.7 cm high and 2.4 cm in diameter, was employed in the five-colligend-system study. Experimental details have been reported previously (*22, 23*). All of the experiments were carried out at constant 1.0×10^{-6} M initial concentration (z_i) of each metal [Analytical Reagent Grade NH_4ReO_4, $(NH_4)_6Mo_7O_{24}$, $(NH_4)_2CrO_4$, Na_2WO_4, and NH_4VO_3], with all five metals always present, and at constant 5.0×10^{-5} M surfactant (hexadecyldimethylbenzylammonium chloride, 97–99% active) concentration. The pH was adjusted with HCl and the ionic strength with NaCl. The initial solution volume was 0.10 liter, the air flow rate was 0.01 liter/min, and the temperature was $22.5 \pm 1.0°C$.

The time dependence of the concentration (z_t) of each metal in the bulk solution was recorded continuously during the experiments by means of radioactive analytical tracers and gamma radiation spectrometry (*22, 23*). At each initial HCl or NaCl concentration, five identical experiments were carried out: in each of the five a different radiotracer was used and thus a different metal oxyanion was monitored, yielding five z_t vs time curves. These curves enabled the calculation of the maximum flotation at foam cease, $1 - z_r/z_i$ (for complete flotation, $z_r = 0$ and $1 - z_r/z_i = 1.0$), and of the first-order flotation rate constant, k, min^{-1} (*20*).

Results

Two distinct series of experiments were carried out, one at constant pH = 6.0 ± 0.2, with the ionic strength varied from 9.0×10^{-5} to 3.0×10^{-1} M with NaCl (*23*); and the other with the pH varied from 4.5 to 0.7

with HCl, and ionic strength variable (22). In both series the Cl⁻ concentration should be expected to play a significant role; the equilibrium position of the ion pair formation reaction between surfactant cations and Cl⁻ should determine the quantity of surfactant cations available to the five metal oxyanions. The cessation of metal oxyanion–surfactant cation ion pair formation should be expected at a certain high Cl⁻ concentration, and thus flotation should cease.

Figures 3 and 4 give the flotation results at pH 6.0 as a function of $p[\text{NaCl}]$, the negative log of the activity of the NaCl in the initial solution. From Fig. 3, at low chloride concentrations, all five metal oxyanions were floated completely, due to the excess surfactant present; at $p[\text{NaCl}] = 3.5$, the residual solution contained only V(V), with the other four metal oxyanions entirely in the foam; and at $p[\text{NaCl}] \leqq 1.5$, Re(VII) and Mo(VI) were separated completely from Cr(VI), W(VI), and V(V). Figure 4 indicates that the order of the first-order flotation rates of the metal oxyanions was similar to that of the maximum flotations $(1 - z_r/z_i)$ in Fig. 3. The flotation rate of Mo(VI) fell off at lower chloride concentrations, indicating that the very efficient flotation of Mo(VI) at higher chloride concentrations was produced by the presence of anionic chloro complexes.

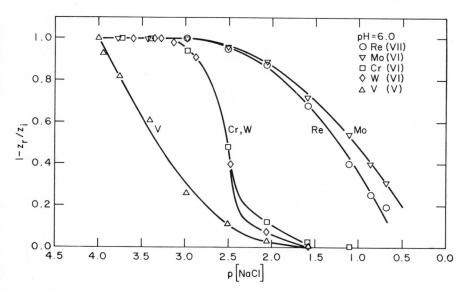

FIG. 3. Maximum flotations of five metal oxyanions vs sodium chloride activity at pH 6.0.

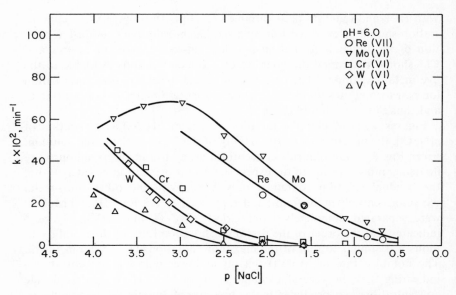

FIG. 4. First-order flotation rate constants of five metal oxyanions vs sodium chloride activity at pH 6.0.

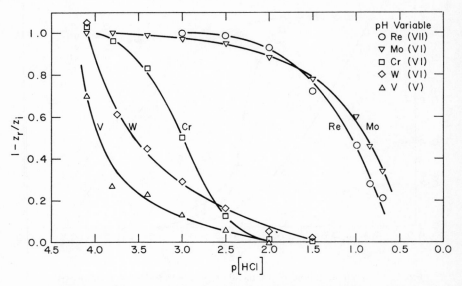

FIG. 5. Maximum flotations of five metal oxyanions in acidic solution vs hydrogen chloride activity.

Figure 5 presents the flotation results at variable, acidic pH (22). The parameter p[HCl] is the negative log of the activity of HCl in the initial solution and was always quite close to the measured pH value. The separations indicated in Figs. 3 and 5 were quite similar, particularly for Re(VII) and Mo(VI). The flotation of Cr(VI) was reduced in the acidic solutions (making the comparison at constant ionic strength, p[HCl] = p[NaCl]), perhaps due to the complete conversion of $CrO_4^{2-} \rightarrow HCrO_4^{-}$. The flotations of W(VI) and V(V) were sharply reduced, probably due to the $WO_4^{2-} \rightarrow HWO_4^{-} \rightarrow H_2WO_4$ and $H_2VO_4^{-} \rightarrow H_3VO_4 \rightarrow VO_2^{+}$ conversions. The role of the transition metal oxyanion species on the flotation behavior is discussed in greater detail below. The effect of p[HCl] on the first-order rate constant (22) was similar to that of p[NaCl], again with reductions in the k values for Cr(VI), W(VI), and V(V) in the acidic solutions, making the comparison at values of p[HCl] = p[NaCl].

For all of the metals in both series of experiments, the collapsed foam volume varied from 2.1 to 0.4% of the initial solution volume of 0.10 liter, providing very rich foams. As would be expected, the foam volume decreased with a decrease in p[HCl] or p[NaCl] (22). For 100% and for 50% flotation of a metal, the ratio of the concentration in the foam to that in the initial solution varied from 50 to 250, and from 25 to 125, respectively, corresponding to the collapsed foam volume values at the extremes of the added electrolyte range.

Discussion

Efforts have been made to predict the selectivity of cationic surfactants for series of anions (14, 19, 24, 61, 62). The extent of interaction is determined by the anion's charge, structure, and degree of hydration. The absolute partial molal entropy of the ion in aqueous solution, $\bar{S}_{298}^{\circ}$ abs, has been suggested as a possible selectivity criterion (24). For ions of like charge, $\bar{S}_{298}^{\circ}$ abs increases with the size of the crystal radius and has been correlated as a decreasing function of the degree of hydration, as measured by the difference between the hydrated radius and the crystal radius (24). An increase in the charge of an ion generally produces a decrease in $\bar{S}_{298}^{\circ}$ abs unless there is a large increase in the crystal radius, and generally produces an increase in the degree of hydration unless there is a large increase in the crystal radius. For two ions of like charge, the less hydrated (largest value of $\bar{S}_{298}^{\circ}$ abs) should be floated preferentially. For two ions of different charges, the more highly charged ion should be floated preferentially unless its entropy is very much smaller that that of the ion of lower charge.

The entropy criterion was tested on a series of anions of like charge (*14*), with values reported in the last column of Table 1; a distinction is made between oxyanions and those not containing oxygen. Some trend is evident, but BrO_3^- is out of order. Perchlorate, with $\bar{S}_{298}^\circ$ abs = 48.7 eu, formed particulates with the cationic surfactant, EHDA–Br, indicating a much stronger interaction than with ClO_3^- (*14*), and $H_2PO_4^-$, with $\bar{S}_{298}^\circ$ abs = 33.5, exhibited a selectivity vs Br^- similar to that of NO_2^- (*63*).

Anion structure may have produced some of the discrepancy in the selectivity vs $\bar{S}_{298}^\circ$ abs correlation attempted in Table 1. For oxyanions of similar structure, a summary is presented in Table 4. For ReO_4^-, MoO_4^{2-}, CrO_4^{2-} and WO_4^{2-}, literature values were used (*64–66*), while for the others, in the absence of reported values, an average was taken of values calculated by the methods of Cobble (*67*) and of Couture and Laidler (*68*). A value of 5.5 eu was used for the proton. For oxyanions with the same charge, again a trend is evident. For Mo(VI), anionic chloro complexes undoubtedly enhanced the flotation; in a sulfuric acid solution with $p[H_2SO_4] = 1.0 \rightarrow 2.0$, ReO_4^- was separated completely from $HMoO_4^-$ by preferential (for ReO_4^-) foam separation with a cationic surfactant (*19*). Diamond (*69*) has reported a selectivity sequence

TABLE 4

Oxyanion Selectivities vs Absolute Partial Molal Entropies

Metal	$1 - Z_r/Z_i$ at $p[NaCl] = 3.0$	$p[HCl] = 3.0$	$k(min^{-1})$ at $p[NaCl] = 3.0$ or $p[HCl]=3.0$	Oxyanion from pK_1 and pK_2		$\bar{S}_{298}^\circ$ abs(eu)
			pH = 6.0			
Re(VII)	0.99		55	ReO_4^-	55	
Mo(VI)	0.99		68	MoO_4^{2-}		25
Cr(VI)	0.96		22	$HCrO_4^-$ (76%)	40	
W(VI)	0.96		16	WO_4^{2-}		26
V(V)	0.31		6	$H_2VO_4^-$ (VO_3^-)	7	
			pH = 3.0			
Re(VII)		0.99	50	ReO_4^-	55	
Mo(VI)		0.97	62	$HMoO_4^-$	45	
Cr(VI)		0.50	14	$HCrO_4^-$	40	
W(VI)		0.29	9	HWO_4^- (H_2WO_4)		
V(V)		0.13	6	$H_3VO_4(HVO_3)$ (VO_2^+)		

TABLE 5

Oxyanion Selectivities at pH 11

Metal	$k(min^{-1})$ in mixture	$k(min^{-1})$ for anion alone	Oxyanion present from pK_1 and pK_2	$\bar{S}_{298}^{\circ}$ abs(eu)
Re(VII)	27	16	ReO_4^-	55
Mo(VI)	6	7	MoO_4^{2-}	25
Cr(VI)	8	5	CrO_4^{2-}	20
W(VI)	5	4	WO_4^{2-}	26
V(V)		2	HVO_4^{2-}	7

$ReO_4^- > CrO_4^{2-} > WO_4^{2-}$ using Dowex–1, a solid anion exchange resin in the Cl^- form.

Additional data were available at pH 11 for a mixture of Re(VII), Mo(VI), Cr(VI), and W(VI), and for each of the five transition metal oxyanions of interest, present by itself in aqueous solution (70). The cationic surfactant was hexadecyltrimethylammonium bromide. Results are presented in Table 5. It would appear that $\bar{S}_{298}^{\circ}$ abs does provide an indication of foam separation selectivity, and thus of oxyanion–surfactant cation interaction, when comparing oxyanions with a wide spread of values of $\bar{S}_{298}^{\circ}$ abs. For oxyanions with similar entropies, the criterion is not of value.

CONCLUSIONS

The cationic surfactant ethylhexadecyldimethylammonium (bromide) is selective for $SCN^- > I^- > ClO_3^- > Br^- > NO_2^-$ as determined from steady-state, equilibrium foam fractionation experiments. The surfactant is modeled as a soluble ion exchanger, and the selectivity coefficients do not vary with the fraction of the exchanger occupied by the preferred ion and are relatively independent of ionic strength.

The cationic surfactant hexadecyldimethylbenzylammonium chloride is selective for Re(VII) $\geq$ Mo(VI) > Cr(VI) $\geq$ W(VI) > V(V) oxyanions in batch, time-dependent foam separation studies from solutions 1.0×10^{-6} M in each metal and containing all five metals. At pH 6.0, with no added NaCl, all five metal oxyanions are floated completely; at $p[NaCl] = 3.5$, V(V) is the only metal remaining in the bulk solution after foam separation; at $p[NaCl] \leq 1.5$, Re(VII) and Mo(VI) can be separated completely from Cr(VI), W(VI), and V(V). Flotation results are similar in acidic solutions, making the comparison at $p[HCl] = p[NaCl]$, except that the flotations of Cr(VI), W(VI), and V(V) are somewhat retarded.

The absolute partial molal entropy in aqueous solution provides an indication of foam separation selectivity by a cationic surfactant when comparing oxyanions with a wide spread of entropy values.

REFERENCES

1. R. A. Lemlich, ed., *Adsorptive Bubble Separation Techniques*, Academic, New York, 1972.
2. F. Sebba, *Ion Flotation*, Elsevier, Amsterdam, 1962.
3. E. Rubin and E. L. Gaden, in *New Chemical Engineering Separation Techniques* (H. M. Schoen, ed.), Wiley-Interscience, New York, 1962, pp. 319–385.
4. I. A. Eldib, in *Advances in Petroleum Chemistry and Refining*, Vol. 7 (K. A. Kobe and J. J. McKetta, Jr., eds.,) Wiley-Interscience, New York, 1962, pp. 66–135.
5. R. A. Lemlich, *Ind. Eng. Chem.*, *60* (10), 16 (1968).
6. B. L. Karger and D. G. DeVivo, *Separ. Sci.*, *3*, 393 (1968).
7. W. Walkowiak and W. Charewicz, *Wiadomosci Chem.*, *25*, 419 (1971).
8. W. Charewicz, *Proceedings of the First Symposium on Rare Elements and Chemical Metallurgy, Karpacz, 1970*, Part II, No. 4(4), Wroclaw, 1971, pp. 97–118.
9. S. F. Kuzkin and A. M. Golman, *Flotatsya Ionov i Molekul*, Atomizdat, Moscow, 1971.
10. W. Charewicz and W. Walkowiak, in *Proceedings of the Eleventh Symposium on Physicochemical Problems of Mineral Processing, Gliwice, 1972*, Vol. 6, Gliwice, 1972, pp. 17–48.
11. Z. Szeglowski, in *Proceedings of the Twelth Symposium on Physicochemical Problems of Mineral Processing, Gliwice, 1973*, Vol. 7, Gliwice, 1973, pp. 121–142.
12. P. Somasundaran, in *Separation and Purification Methods*, Vol. 1 (E. S. Perry and C. J. van Oss, eds.), Dekker, New York, 1973, pp. 117–198.
13. R. B. Grieves, D. Bhattacharyya, and P. J. W. The, *Can. J. Chem. Eng.*, *51*, 173 (1973).
14. R. B. Grieves and P. J. W. The, *J. Inorg. Nucl. Chem.*, *36*, 1391 (1974).
15. W. Charewicz and J. Niemiec, *Nukleonika*, *14*, 17 (1969).
16. W. Charewicz and J. Niemiec, *Ibid.*, *14*, 607 (1969).
17. W. Charewicz, in *Proceedings of the First Symposium on Rare Elements and Chemical Metallurgy, Karpacz, 1970*, Part II, No. 4(4), Wroclaw, 1971, pp. 71–88.
18. W. Charewicz, *Rocz. Chem.*, *46*, 1979 (1972).
19. W. Charewicz and W. Walkowiak, *Separ. Sci.*, *7*, 631 (1972).
20. W. Charewicz, *J. Appl. Chem. Biotechnol.*, *23*, 743 (1973).
21. W. Charewicz, *Nukleonika*, In Press.
22. W. Charewicz and R. B. Grieves, *J. Inorg. Nucl. Chem.*, *36*, 2371 (1974).
23. W. Charewicz and R. B. Grieves, *Anal. Lett.*, *7*, 233 (1974).
24. R. B. Grieves and D. Bhattacharyya, *Ibid.*, *4*, 603 (1971).
25. W. Charewicz and J. Niemiec, *Third National Symposium on Application of Isotopes in Technique*, Report No. 48, Szczecin, 1966.
26. W. Charewicz and J. Niemiec, *Neue Hütte*, *13*, 401 (1968).
27. W. Charewicz and J. Niemiec, *Nukleonika*, *14*, 799 (1969).
28. W. Charewicz and W. Walkowiak, in *Proceedings of the Eleventh Symposium on Physicochemical Problems of Mineral Processing, Gliwice, 1972*, Vol. 6, Gliwice, 1972, pp. 49–56.

29. W. Walkowiak, in *Proceedings of the First Symposium on Rare Elements and Chemical Metallurgy, Karpacz, 1970*, Part II, No. 4(4), Wroclaw, 1971, pp. 89–96.
30. W. Walkowiak and A. Bartecki, *Nukleonika, 18*, 209 (1973).
31. W. Charewicz and T. Gendolla, *Chem. Stosow., 16*, 383 (1972).
32. W. Walkowiak and A. Bartecki, *Nukleonika, 18*, 133 (1973).
33. W. Walkowiak, in *Proceedings of the Second Symposium on Rare Elements and Chemical Metallurgy, Karpacz, 1972*, Part II, No. 17(2), Wroclaw, 1973, pp. 433–442.
34. K. Jurkiewicz and A. Waksmundzki, *Rocz. Chem., 47*, 1457 (1973).
35. M. Bittner, J. Mikulski, and Z. Szeglowski, *Nukleonika, 12*, 599 (1967).
36. Z. Szeglowski, M. Bittner, J. Mikulski, and T. Machaj, in *Proceedings of the Second Symposium on Rare Elements and Chemical Metallurgy, Karpacz, 1972*, Part II, No. 17(2), Wroclaw, 1973, pp. 459–464.
37. Z. Szeglowski, M. Bittner, and J. Mikulski, in *Proceedings of the Second Symposium on Rare Elements and Chemical Metallurgy, Karpacz, 1972*, Part II, No. 17(2), Wroclaw, 1973, pp. 465–472.
38. W. Charewicz, T. Gendolla, and D. Podgorska, in *Proceedings of the Second Symposium on Rare Elements and Chemical Metallurgy, Karpacz, 1972*, Part II, No. 17(2), Wroclaw, 1973, pp. 273–284.
39. *Stability Constants*, Special Publication No. 17, The Chemical Society, London, 1964.
40. A. Carrington and M. C. R. Symons, *Chem. Rev., 63, 443* (1963).
41. N. Bailey, A. Carrington, K. A. K. Lott, and M. C. R. Symons, *J. Chem. Soc., 1960*, 290.
42. K. B. Yatsimirskii and I. I. Alekseeva, *Izv. Vyssh. Ucheb. Zaved., Khim. Khim. Tekhnol., 1*, 53 (1958).
43. C. P. Vorobev, I. P. Davydov, and T. W. Shilin, *Zh. Neorg. Khim., 12*, 2142 (1967).
44. L. G. Sillen, *Quart. Rev.* (London), *13*, 416 (1959).
45. Y. Sasaki, I. Lindqvist, and L. G. Sillen, *J. Inorg. Nucl. Chem., 9*, 93 (1959).
46. A. K. Babko and B. I. Nabivantes, *Zh. Neorg. Khim., 2*, 2085 (1957); *2*, 2096 (1957).
47. J. Ying-Peh Tong and E. L. King, *J. Amer. Chem. Soc., 75*, 6180 (1953).
48. G. P. Haight, Jr., D. C. Richardson, and N. C. Coburn, *Inorg. Chem., 3*, 1777 (1964).
49. J. Hala, O. Navratil, and V. Nechuta, *J. Inorg. Nucl. Chem., 28*, 553 (1966).
50. G. Schwarzenbach, G. Geier, and J. Littler, *Helv. Chim. Acta., 45*, 2601 (1962).
51. K. B. Yatsimirskii and V. F. Romanov, *Zh. Neorg. Khim., 10*, 1607 (1965).
52. J. Chojnacka, *Nukleonika, 12*, 729 (1967).
53. A. Schwarzenbach and A. Geier, *Helv. Chim. Acta., 46*, 906 (1963).
54. D. Dyrssen and T. Sekine, *J. Inorg. Nucl. Chem., 26*, 981 (1964).
55. M. T. Pope and B. W. Dale, *Quart. Rev.* (London), *4*, 527 (1968).
56. N. Ingri and F. Brito, *Acta. Chem. Scand., 13*, 1971 (1959).
57. R. M. Diamond, *J. Phys. Chem., 61*, 75 (1957).
58. G. Carpeni, *Bull. Soc. Chim. Fr., 14*, 496 (1947).
59. A. Bartecki and J. Kaminski, *Rocz. Chem., 44*, 1839 (1970); *45*, 315 (1971).
60. F. J. C. Rossotti and H. Rossotti, *Acta. Chem. Scand., 10*, 957 (1956).
61. E. Rubin and J. Jorne, *Separ. Sci., 4*, 313 (1969).
62. K. Shinoda and M. Fujihira, *Advan. Chem. Series, 79*, 198 (1968).

63. R. B. Grieves and D. Bhattacharyya, *Separ. Sci.*, *1*, 81 (1966).
64. R. W. Gurney, *Ionic Processes in Solution*, McGraw-Hill, New York, 1953, p. 267.
65. F. D. Rossini, D. D. Wagman, W. H. Evans, S. Levine, and J. Jaffe, *Selected Values of Chemical Thermodynamic Properties*, National Bureau of Standards, Circular 500, Washington, 1952.
66. W. M. Latimer, *The Oxidation States of the Elements and Their Potentials in Aqueous Solutions*, Prentice-Hall, New York, 1952.
67. J. W. Cobble, *J. Chem. Phys.*, *21*, 1443, 1446, (1953).
68. A. M. Couture and K. J. Laidler, *Can. J. Chem.*, *35*, 202 (1957).
69. R. M. Diamond and D. C. Whitney, in *Ion Exchange*, Vol. 1 (J. A. Marinsky, ed.), Dekker, New York, 1966, p. 302.
70. W. Charewicz, To Be Published.

Separation Using Foaming Techniques

P. SOMASUNDARAN

HENRY KRUMB SCHOOL OF MINES
COLUMBIA UNIVERSITY
NEW YORK, NEW YORK 10027

Abstract

Separation of various chemical components from each other is often the most difficult step in analytical procedures. The problems attached to separation become further magnified when the species concentations are extremely low. A group of techniques that has proven useful especially in dilute solutions for separating and concentrating metallic as well as nonmetallic ions and complexes, proteins, microorganisms, particulates, etc. is the adsorptive bubble separation techniques. Minerals have indeed been treated using some of these techniques for decades. The success of these processes is primarily dependent upon differences in the natural surface activity of various species or particulates in the system or in their tendency to associate with surfactants. The efficiency of the process is determined by such variables as solution pH, ionic strength, concentration of various activating and depressing agents, and temperature. A proper control of variables offers an opportunity to separate a variety of metallic and nonmetallic species and particulates. In this paper the principles governing various foam separation techniques, particularly froth flotation, are presented along with the recent results on the role of variables that can be controlled to achieve complete removal of species and particulates for analytical purposes.

INTRODUCTION

A variety of materials can be concentrated as well as separated from one another using foam separation techniques that make use of the tendency of surface-active components in a solution to preferentially concentrate at the solution/gas interface. Nonsurface active agents that are capable of associating with surface-active agents can also be separated using these

117

techniques. A comprehensive list of materials that have been treated using foam separation techniques is available in a recent publication (*1*). The list includes various anions such as alkylbenzyl sulfonate; chromate; cyanide and phenolate; cations of, for example, dodecylamine, mercury, lead, and strontium; proteins; microorganisms; and minerals. The attractive feature of this group of techniques is its effectiveness in the concentration range that is too dilute for the successful use of most other techniques. Furthermore, these techniques are ideally suitable for also treating materials that are too sensitive to changes in temperature. In this paper first principles that govern the separation will be briefly described and then the role of common variables that can be controlled for optimizing the separation will be examined with the help of typical examples.

Various foam separation methods studied in the past are listed in Table 1 on the basis of (a) the particle size of the materials and (b) the mechanism by which they are separated. Thus we have foam fractionation for separating surface-active species such as detergents from aqueous solution, and molecular and ion flotation for the separation of nonsurfactive species such as strontium, lead, and cyanides that can be made to associate with various surfactants. Nonsurface-active species that are separated in this manner are called colligends, and the surface-active agents used to separate them are called collectors. The separation of microscopic size organisms and proteins which are naturally surface active has been called foam flotation,

TABLE 1

Various Adsorptive Bubble Separation Methods Classified on the Basis of
Mechanism of Separation and Size of the Material Separated (*1*)

Mechanism	Size range		
	Molecular	Microscopic	Macroscopic
Its natural surface activity	Foam fractionation; example, detergents from aqueous solutions	Foam flotation; examples, microorganisms, proteins, dyes	Froth flotation of nonpolar minerals; example, sulfur
Its association with surface active species	Ion flotation, molecular flotation; examples, Sr^{2+}, Ag^{2+}, Pb^{2+}, Hg^{2+}, cyanides, phosphates	Microflotation; examples, particulates in waste, microorganisms	Froth flotation; example, minerals such as silica. Precipitate flotation (1st and 2nd kind); example, ferric hydroxide

and that of subsieve size particulates which are not surface active by themselves has been called microflotation. Froth flotation, used in the mineral beneficiation area for the last 60 years, refers to the separation of sieve-size particulates. It must be noted that, as opposed to all other foam separation techniques, froth flotation employs a relatively high gas flow-rate under turbulent conditions. Next to froth flotation, the most useful foam separation technique is precipitate flotation where the species to be separated is first precipitated, usually by a change of solution pH, and then floated with the help of surfactants which adsorb on the precipitates. In addition to the above techniques, there are also certain nonfoaming separation methods such as bubble fractionation (2) and solvent sublation (3) where adsorption at interfaces is again the basis for concentration, but the adsorbed material is collected for removal in another liquid that is immiscible with the bulk solution. All the above foaming and nonfoaming methods have been collectively called adsorptive bubble separation techniques (4).

METHODS

In practice, foam separation consists of aeration at a low flow-rate of the solution containing the species to be separated and a surfactant, if the species are not naturally surface active, and separation of the adsorbed components by simply removing the foam mechanically and breaking it using various chemical, thermal, or mechanical methods (5). Recovery or percent removal and the grade of the product is increased by using a stripping mode and an enriching mode, respectively. In the former mode the feed is introduced into the foam so that some amount of separation takes place while it is descending through the foam itself. In the enriching mode a certain amount of reflux for increasing the separation factor is achieved by feeding part of the foamate back to the top of the column. In a conventional froth flotation cell, a pulp of particles containing appropriate reagents is agitated using an impeller, and air is sucked in or sometimes fed into the cell. Those particles that are hydrophobic or that have acquired hydrophobicity adhere to air bubbles and thus rise to the cell top where they are removed by skimming. A flotation cell that is suitable for analytical purposes is the modified Hallimond cell (6) shown in Fig. 1. It consists mainly of two parts, a glass well with a fritted glass disk at the bottom connected to a gas reservoir and a bent top part with a stem just above the bend. Stirring can be accomplished either by means of a magnetic stirrer or by means of an impeller. Gas flow-rate is controlled

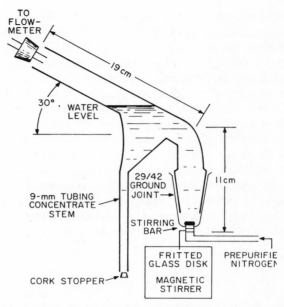

FIG. 1. Modified Hallimond cell for laboratory flotation (6).

by adjusting the pressure in the gas reservoir. Hydrophobic species will collect in the cell top and in the stem and can be separated from those in the well by decantation followed by rinsing of the cell top.

PRINCIPLES

Foam separation is based on the adsorption of surfactants at the liquid/air interface and the association of various chemical species and particulates with these surfactants. Surfactant adsorption at the liquid/air interface takes place because interaction energy between the nonpolar hydrocarbon chains of the surfactant and the polar water molecules is less than the interaction energy between water molecules themselves and therefore the presence of the organic molecules in bulk water is energetically less favorable than their presence out of the bulk water at the interface. As the size of the nonpolar chain increases, it becomes more and more energetically unfavorable for the chains to stay in the bulk water. An increase in chain length therefore causes an increase in its adsorption at the liquid/air interface (7) and therefore its percent removal by foaming. If the number of polar groups or the number of double and triple bonds are increased, however, the adsorption and consequently the separation can be expected to be poorer.

It must be noted that the adsorption density of a surfactant will not increase by any significant amount on increasing its bulk concentration above its critical micelle concentration. The ratio of the surfactant concentration at the surface to that in the bulk, known as the distribution factor, which is actually a measure of the possible separation, will therefore decrease above the critical micelle concentration. Efficiency of separations can therefore be expected to be higher in dilute solutions than in concentrated solutions. Of course, there must be enough surfactant of one type or another to produce foams. Experimental results of Aenlle (8) for the distribution factor of a surfactant Aresket as well as of a uranyl species that was separated using Aresket did, in fact, show the factor to be largest in dilute Aresket solutions.

If the aim is to separate one surfactant from another or to purify a surfactant, it is most important to conduct the separation below the various critical micelle concentrations. Since micelles of a surfactant solubilize other surfactants and thereby reduce their adsorption at the liquid/gas interface, their distribution factor will be lower above the critical micelle concentration of any surfactant than below it. Foaming for purification has to be conducted, therefore, below the concentration at which micelle formation takes place.

The extent of separation of species is also dependent upon the removal of the bulk solution from the foams before the collection of the foams. The foam must therefore drain as much as possible without rupturing. This is dependent on various properties such as viscosity of the bulk liquid, electrical double-layer repulsion between the two surfaces of adjoining bubbles, and the elasticity and viscosity of the surface film. The role of these factors in foam separation techniques has been discussed by Lemlich et al. (9).

Foam separation of nonsurface-active materials is, as mentioned earlier, dependent upon their association with surface-active materials. This association can arise from chemical interaction between the two species, which often leads to the formation of complexes, or from physical interaction. Ion flotation, for example, is based on the association between the ions and the oppositely charged surfactant species due to electrostatic attraction.

Flotation of quartz using alkylamines is another example where electrostatic interaction is put to use for separation as well as concentration.

In this case quartz is negatively charged and hence adsorbs cationic aminium ions and thereby acquires hydrophobicity. Froth flotation of a large number of sulfide minerals, on the other hand, depends on the chemical reaction of the surfactant with the surface species of the mineral. Flotation of galena (PbS) using potassium xanthate is an example. Flotation based on the chemisorption of surfactants is employed for separation of several oxide and salt-type minerals in addition to sulfides. To enhance the adsorption of surfactants on desired minerals or to prevent the adsorption on others, a variety of reagents known as modifiers are used in practice (10). They include alkali or acid to adjust the pH; alkali sulfides, cyanides, and sulfites to depress the flotation of certain sulfide minerals; copper sulfate to activate the flotation of zinc sulfide; ferric chloride, calcium chloride, and cupric nitrate to activate the flotation of quartz; and polymeric reagents such as starch to depress the flotation of various salt-type minerals.

EFFECT OF VARIABLES

For the successful use of foam separation techniques as analytical tools, it is necessary to achieve, as much as possible, complete separation of the species of interest. Toward this goal, the effect of all relevant controllable variables on percent removal and selectivity by foam separation methods will be examined.

Chain Length of the Surfactant

For reasons given earlier, an increase in length of the nonpolar part of the surfactant should lead to an increase in its adsorption at interfaces and therefore its separation. Our results (7) for alkylammonium acetates have in fact shown that these reagents adsorb at the solution/air interface in increasing quantities as the chain length is increased. The length of the hydrocarbon chain of the surfactant was found to affect froth flotation of materials in a similar manner (11). Hallimond tube flotation of quartz with alkylammonium acetates of varying chain length is shown in Fig. 2 as a function of the surfactant concentration. It can be seen that the percent removal increases drastically on increasing the chain length of the surfactant. The chain length effect on flotation was ascribed to the tendency of the longer chains to associate into two dimensional aggregates called hemi-micelles at the solid–liquid interface (11, 12). The driving force for this association is the cohesive interaction between the chains and therefore is dependent on the chain length.

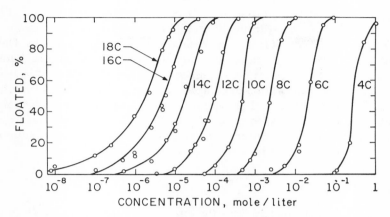

FIG. 2. The effect of hydrocarbon chain length on the flotation of quartz in alkylammonium acetate solutions (*11*).

Surfactant Concentration

It is also evident from Fig. 2 that the percent removal is strongly dependent on the concentration of the surfactant. This is not due to the dependence of physical properties or stability of foam since effects of such factors are eliminated in a Hallimond tube test. Rubin and co-workers (*13*), among others, have also observed dependence of collector concentration on the precipitate flotation of copper species. They found a collector to colligend ratio of one to be necessary in their case to get nearly complete removal of the copper. Concentration of the collector was found to be even more critical if ion flotation is used instead of precipitate flotation. Rubin and Lapp (*14*) have reported that while 100% removal of zinc species is possible using a collector to colligend ratio of 0.2 in the pH range of 8 to 11 where zinc hydroxide precipitates, almost no flotation is obtained at that ratio below pH 8 when zinc is present in dissolved ionic form. A larger quantity of collector was needed to remove the zinc completely under these conditions.

A great excess of collector has, however, been found to reduce the flotation of minerals (*15*), precipitates (*16*), and ions (*17*). In the case of particulate flotation, this is sometimes due to a reduction in the size of the bubbles to such a level that the bubbles could not levitate the large number of particles that collect on them (*15*). Adsorption of a second layer of collector at higher concentration with an orientation opposite to that of the first layer or adsorption of micelles can also cause a decrease in flotation, but less likely to do so in most cases since only a small fraction of the surface needs to be hydrophobic for flotation to occur. The inhibitive effect of excess collector on ion flotation has been proposed by Davis and Sebba (*17*) to be mainly due to possible crowding of the bubble surface by the collector ions themselves and the formation of micelles with consequent adsorption of colligend ions on the nonfloatable micelles. Both for ion flotation and precipitate flotation, an optimum collector to colligend ratio is reported to exist (*18–20*). One technique reported by Grieves et al. (*21, 22*) for removing additional colligend from solution is by adding the collector in pulses instead of in one dose.

Solution pH

The role of pH in determining the form of the species present in solution and thereby its flotation as ions or precipitates is evident in the work of Rubin and Lapp (*14*) discussed earlier. The effect of pH on particulate separation is even more significant. In fact, it is primarily the proper choice of pH along with the type of collector that enables one to selectively float one type of particulate from another and thus obtain their separation. This effect of pH is illustrated in Fig. 3 where flotation of calcite with an anionic and a cationic collector is given as a function of pH at two con-

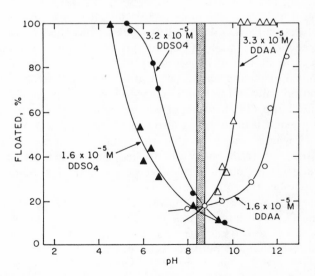

FIG. 3. The effect of pH on the anionic and cationic flotation of calcite (*23*).

centrations (23). The isoelectric point of calcite as measured by streaming potential is about pH 8 to 9.5 (23). It can be seen that significant flotation with an anionic collector is possible only below the isoelectric point where the particles are positively charged. Similarly flotation with a cationic collector is possible only above the isoelectric point where the particles are negatively charged. For analytical purposes, one is interested in determining how this material can be separated if it is mixed with another material, for example, quartz. The isoelectric point of quartz is near pH 2. Between pH 2 and 8 quartz is therefore negatively charged while calcite is positively charged. It is therefore possible to achieve a separation either by floating calcite with an anionic collector at, for example, pH 7 or by floating quartz with a cationic collector at that pH. It must be noted that this separation is likely to fail at much lower pH values since dissolved calcium species, for reasons given elsewhere, will activate the anionic flotation and depress the cationic flotation of quartz. Control of pH can be similarly used for separation purposes with other foam techniques. As an example, distribution factors obtained by Karger et al. (24) for mercury and iron in the presence of an amine are given in Fig. 4. These results suggest that the separation of mercury from iron can be obtained either by floating the former at higher pH values or the latter at lower pH values.

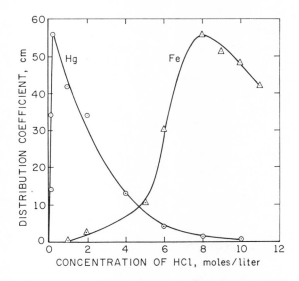

FIG. 4. Distribution coefficients for Fe and Hg as a function of HCl concentration in the presence of a cationic surfactant (24).

In addition to the above effects, pH also influences separation due to dependence of the collector hydrolysis on it. A typical example of this is the cationic flotation of quartz in basic solutions. Quartz is negatively charged above pH 2 and therefore it should be possible to float it with a cationic collector above this pH. In practice, however, one gets very little flotation of quartz with dodecylammonium acetate above pH 12 (25). Above pH 12, most of this collector is in its neutral molecular form and under such conditions it is apparently unable to adsorb on quartz and make it hydrophobic. Neutral molecules can, however, act as good collectors when present along with ionic surfactant species. Total adsorption of the surfactant at the solid/solution interface and hence flotation in a system containing both ionic and neutral surfactant species appears from the experimental results to be higher than when the same amount of surfactant is present totally in one or the other form. This is suggested to be due to the fact that, if some of the species adsorbed on the solid are neutral, they can actually screen the repulsion between the charged heads of the adsorbed ions. Based on the same principle, Fuerstenau and Yamada (26) were able to enhance flotation by adding long-chain alcohol along with the collector.

Ionic Strength

It is possible to float quartz with an amine collector above pH 2 because, as mentioned earlier, cationic aminium species adsorb electrostatically on the negatively charged quartz. Such electrostatic adsorption of aminium ions will take place in competition with other ions that are similarly charged. A significant increase in concentration of nonsurfactive cations will therefore decrease the adsorption of the cationic collector ions on the solid and hence also its flotation. Results obtained recently for the cationic flotation of quartz show this to be the case (see Fig. 5). Potassium nitrate thus acts as a depressant for the cationic flotation of quartz. Modi and Fuerstenau (28) have observed similar effects of sodium chloride on the anionic flotation of alumina. When sodium sulfate was added to the system instead of sodium chloride, the depression of flotation was even larger. This larger effect of sulfate over that of chloride results from the tendency of the bivalent sulfate to strongly adsorb and compete with dodecylsulfate more than the monovalent chloride. The above effect can also be used to enhance the flotation of a particle that has a charge opposite to that of the collector. Modi and Fuerstenau (28) were thus able to get complete flota-

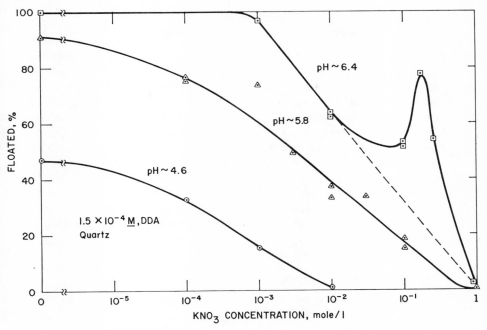

FIG. 5. Effect of ionic strength on the flotation of quartz with dodecylammonium acetate (DDA) (27).

tion of the positively charged alumina at pH 6 using a cationic surfactant
by adding sufficient sodium sulfate to get a concentration of 10^{-2} mole/
liter. Flotation of alumina occurs in the presence of bivalent sulfate ions
because specifically adsorbing sulfate at such concentrations reverses the
charge of the alumina particle to a negative sign and thus makes the adsorption of cationic surfactant possible. Similarly, negatively charged particles
can be floated with anionic collectors if the particles are first modified by
means of cations such as calcium and magnesium. These ions can often
function most effectively in the pH range where they are in the hydrolyzed
soluble form. Fuerstenau et al. (29) studied the role of iron, aluminum,
lead, manganese, magnesium, and calcium in the anionic flotation of
quartz as a function of pH and found that each cation began to function
as an activator as the metal ions began to hydrolyze and ceased function-
ing in that manner when the solution pH corresponded to that at which
the metal hydroxides begin to precipitate.

Concentration of Complexing Ions

Certain metallic species are first separated by complexing them with inorganic agents and then floating them with a collector. The concentration of the species used for complexing is found to be critical in several cases. Anionic flotation of copper hexacyanoferrate is reported to be impossible when the concentration of Cu^{2+} ions is below the point of stoichiometry equivalence (16). Shakir (30) also observed the concentration of the complexing anion to be critical for complete separation for the flotation of uranyl carbonate or uranyl bicarbonate anionic complex with cetyltrimethylammonium bromide. Whereas nearly 100% removal of the uranium could be obtained from 10^{-4} mole/liter uranyl solution in the presence of 0.01 M carbonate or bicarbonate, only 25% of it could be removed when the carbonate or bicarbonate concentration was 0.4 mole/liter. Evidently, control of the concentration of the complexing ions is important to get maximum recovery.

Flocculants and Polymers

In several cases, complete flotation can be obtained by adding auxiliary reagents, particularly flocculants. Flotation of *B. cereus* and illite with sodium lauryl sulfate was increased to almost 100% by Rubin et al. by adding alum (31, 32). Selective flocculation followed by flotation has proven to be a successful method for the separation of fine hematite from quartz (33). In such processes, the grade of the product will indeed suffer if the flocculant is not selective.

Separation by flotation is markedly affected by polymeric-type reagents such as starch. Figure 6 illustrates the effect of starch addition on the oleate flotation of calcite (34). It can be seen that small additions of starch decreases the flotation drastically. It is interesting to note that, unlike most other depressants, starch does not reduce flotation by inhibiting adsorption of surfactant on the colligend particles. In fact, adsorption of oleate on calcite was found to be higher in the presence of starch than in its absence. Thus even though the particles adsorbed more surfactant in the presence of starch, the particles are hydrophilic. This effect was ascribed to the peculiar helical structure that starch assumes in the presence of hydrophobic species or under alkaline conditions. The structure of helix is such that its interior is hydrophobic and its exterior is hydrophilic. Mutual enhancement of adsorption is possibly due to the formation of a helical starch–oleate clathrate with the hydrophobic oleate held inside

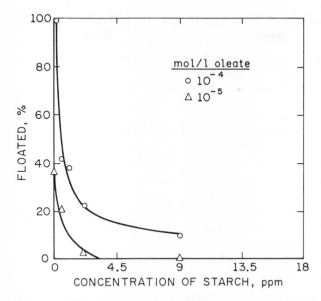

Fɪɢ. 6. Effect of addition of starch on the flotation of calcite using oleate (34).

the hydrophobic starch interior. The hydrophilic nature of calcite in the presence of starch and oleate results from the fact that the adsorbed oleate is obscured from solution by such wrapping by starch helixes whose exterior is hydrophilic and also by simple overwhelming by massive starch species.

Physical Variables

Among the physical variables, flow rate of the gas, bubble size distribution, agitation, etc. do not produce any primary effects on the ultimate total flotation of materials, even though at extreme minimum and maximum levels some of these factors can influence the extent of removal. For example, excessive aeration could conceivably produce enough turbulence in the system to reduce the separation by molecular or ion flotation. The rate of removal of colligend is indeed dependent upon the levels of the above physical variables. One variable that has been found to be important, at least in froth flotation, is the temperature of the solution. In the case of flotation dependent on the physical adsorption of collector, the recovery can be expected to decrease with increasing temperature, and in the case of that dependent on chemisorption, it can be expected to increase with it. There has been very little actual experimental work on temperature effects on froth flotation except to show that flotation of hematite with oleate increases with the temperature at which the material is reagentized. Our recent results (35) have however shown this to be true only at low ionic strength conditions. Above 2×10^{-3} N ionic stength, flotation in our case is found to markedly decrease with increasing reagentizing temperature. The effect of temperature on foam separation has been reported by Rubin et al. (13) and Schoen and Mazzella (36) to be very little. Sheiham and Pinfold (20) noted an increase in the rate of removal of strontium carbonate precipitate on increasing the flotation temperature from 15 to 30°C. In the case of solvent sublation of hexacyanoferrate, however, Spargo and Pinfold (37) noted a marked decrease in recovery on increasing the temperature. In the foam separation techniques, the effect of temperature on foam drainage and stability will also play a role in determining the separation efficiency.

EXAMPLES OF FOAM SEPARATED MATERIALS

A listing of typical materials separated by various foaming techniques is given below. For a more detailed list or the details of conditions of separation, a previous publication (1) or the original reference must be consulted.

Foam Fractionation. Alkylbenzenesulfonate (38, 39), amines (40), fatty acids (41), alcohols (41), detergents from waste waters (42), and surfactants from paper and pulp mill streams (43).

Ion Flotation. Ag (40), Au (44), Be (36), Co (45), Cu (46), Fe (36), Ra (36), cyanide (47), dichromate (48), and phenolate (48). See also *Ion Flotation* by Sebba (49).

Foam Flotation. Albumin (50), hemoglobin (51), catalase (52), algae (53), and methylcellulose (54).

Microflotation. Kaolinite (55), iron dust (56), deactivated carbon (55), *E. coli* (57), Ce (58), U (59), and trace elements in seawater (60).

Precipitate Flotation. Silver (61), chromium (62), copper (63), strontium (63), and zinc (64).

Froth Flotation. Sulfur (65), coal (66), calcium phosphate (67), feldspar (65), potassium chloride (68), and wastes (69). See also *Froth Flotation, 50th Anniversary Volume* (70).

CONCLUSIONS

The above discussion shows that complete removal of materials from solutions can be achieved by optimizing the levels of certain operating variables of foam separation techniques. In fact, these techniques should prove even more useful for analytical purposes than for the conventional industrial processes since in the former case it is possible to change pH, concentration of collector, etc. as much as necessary without any attention to the economic aspects of the processes. Furthermore, flotation times that are longer than that used in industrial operations can be employed in an analytical procedure in order to achieve complete recovery. Nonfoaming adsorptive bubble separation techniques also appear promising for use as analytical separation tools. For example, solvent sublation, where the colligend is levitated by means of bubbles into a thin solvent layer above the solution, is reported to be capable of being more efficient than solvent extraction (71) since it is possible to attain colligend concentrations in the solvent above the equilibrium value with the former technique.

Because foam separation techniques can concentrate from solutions that are as dilute as 10^{-10} mole/liter, they will find applications as concentration techniques. However, their high degree of selectivity suggests that they can also be used for the separation of one material from another. Few other techniques can actually concentrate or separate from solutions containing parts per billion of the species as efficiently as the foam separation techniques.

REFERENCES

1. P. Somasundaran, "Foam Separation Methods," in *Separation and Purification Methods*, Vol. 1 (E. S. Perry and C. J. van Oss, eds.), Dekker, New York, 1972, p. 117.
2. R. Lemlich, "Principles of Foam Fractionation," in *Progress in Separation and Purification*, Vol. 1 (E. S. Perry, ed.), Wiley-Interscience, New York, 1968, pp. 1–56. R. Lemlich, *Ind. Eng. Chem.*, 60, 16 (1968). S. Bruin, J. E. Hudson, and A. I. Morgan, Jr., *Ind. Eng. Chem., Fundam.*, 11, 175 (1972).
3. B. L. Karger, T. A. Pinfold, and S. E. Palmer, *Separ. Sci.*, 5, 603 (1970).
4. R. Lemlich, ed., *Adsorptive Bubble Separation Techniques*, Academic, New York, 1972, p. 1.
5. M. Goldberg and E. Rubin, *Ind. Eng. Chem., Process. Des. Develop.*, 6, 195 (1967).
6. D. W. Fuerstenau, P. H. Metzger, and G. D. Seele, *Eng. Mining J.*, 158, 93 (1957).
7. P. Somasundaran, *Trans. AIME*, 241, 105 (1968).

8. E. O. Aenlle, *An. Real Soc. Espan. Fis. Quim.*, *42*, 179 (1946); *Chem. Abstr.*, *41*, 4649i (1947).

9. R. A. Leonard and R. Lemlich, *Amer. Inst. Chem. Eng. J.*, *11*, 18 (1965). F. Shih and R. Lemlich, *Ibid.*, *13*, 751, (1967). R. Lemlich, "Principles of Foam Fractionation and Drainage," in Ref. *4*, pp. 33–50.

10. A. M. Gaudin, *Flotation*, McGraw-Hill, New York, 1957, pp. 282–326

11. D. W. Fuerstenau, T. W. Healy, and P. Somasundaran, *Trans. AIME*, *229*, 321 (1964).

12. P. Somasundaran and D. W. Fuerstenau, *J. Phys. Chem.*, *70*, 90 (1966).

13. A. J. Rubin, J. D. Johnson, and J. C. Lamb, *Ind. Eng. Chem.*, *Process Des. Develop.*, *5*, 368 (1966).

14. A. J. Rubin and W. L. Lapp, *Separ. Sci.*, *6*, 357 (1971).

15. P. Somasundaran and R. D. Kulkarni, *Trans I.M.M.*, *82*, C164–167 (1973).

16. T. A. Pinfold, in Ref. *4*, p. 79.

17. B. M. Davis and F. Sebba, *J. Appl. Chem.*, *16*, 293 (1966).

18. M. W. Rose and F. Sebba, *Ibid.*, *19*, 185 (1969).

19. N. Aoki and T. Sasaki, *Bull. Chem. Soc. Japan*, *39*, 939 (1966).

20. I. Sheiham and T. A. Pinfold, *J. Appl. Chem.*, *18*, 217 (1968).

21. R. B. Grieves and T. E. Wilson, *Nature*, *205*, 1066 (1965).

22. R. B. Grieves and D. Bhattacharya, *J. Appl. Chem.*, *19*, 115 (1969).

23. P. Somasundaran and G. E. Agar, *J. Colloid Interfac. Sci.*, *24*, 433 (1967).

24. P. L. Karger, R. P. Poncha, and M. W. Miller, *Anal. Lett.*, *1*, 437 (1968).

25. P. Somasundaran and D. W. Fuerstenau, *Trans. AIME*, *241*, 102 (1968).

26. D. W. Fuerstenau and B. J. Yamada, *Ibid.*, *223*, 50 (1962).

27. P. Somasundaran, *Ibid.*, *255*, 64 (1974).

28. H. J. Modi and D. W. Fuerstenau, *Ibid.*, *217*, 381 (1960).

29. M. C. Fuerstenau, C. C. Martin, and R. B. Bhappu, *Ibid.*, *226*, 449 (1963).

30. K. Shakir, *J. Appl. Chem. Biotechnol.*, *23*, 339 (1973).

31. A. J. Rubin, in Ref. *4*, p. 216

32. A. J. Rubin and S. C. Lackey, *J. Amer. Water Works Assoc.*, *60*, 1156 (1968).

33. R. Sisselman, *Mining Eng.*, *25*, 45 (1973).

34. P. Somasundaran, *J. Colloid. Interfac. Sci.*, *31*, 557 (1969).

35. P. Somasundaran and R. D. Kulkarni, "Conditioning Temperature–Ionic Strength Interactions in Hematite Flotation," 103rd Annual AIME Meeting, Dallas, February 1974.

36. H. M. Schoen and G. Mazzella, *Ind. Water Wastes*, *6*, 71 (1961).

37. P. E. Spargo and T. A. Pinfold, *Separ. Sci.*, *5*, 619 (1970).

38. C. A. Brunner and D. G. Stephan, *Ind. Eng. Chem.*, *57*, 40 (1965).

39. P. H. McGauhey and S. A. Klein, *Public Works*, *92*, 101 (May 1961).

40. C. Walling, E. E. Ruff, and J. L. Thorton, Jr., *J. Phys. Chem.*, *56*, 989 (1952).

41. E. Rubin and E. L. Gaden, Jr., in *New Chemical Engineering Techniques*, (H. M. Schoen, ed.), Wiley-Interscience, New York, 1962, pp. 319–385.

42. W. Krygielowa, J. Kucharski, and J. Wasowiczowa, *Chem. Abstr.*, *68*, 15871n (1968).

43. J. L. Rose and J. F. Sebald, *Tappi*, *51*, 314 (1968).

44. L. Dobresco and V. Dobresco, *Rev. Minelor* (Bucharest), *19*, 231 (1968).

45. B. L. Karger and M. W. Miller, *Anal. Chim. Acta*, *48*, 273 (1969).

46. A. J. Rubin and J. D. Johnson, *Anal. Chem.*, *39*, 298 (1967).

47. R. B. Grieves and D. Bhattacharya, *Separ. Sci.*, *4*, 301 (1969).
48. R. B. Grieves, *Ibid.*, *1*, 395 (1966).
49. F. Sebba, *Ion Flotation*, Elsevier, New York, 1962.
50. W. W. Eckenfelder, Jr. and D. J. O'Connor, *Biological Waste Treatment*, Pergamon, New York, 1961.
51. F. Schutz, *Nature*, *139*, 629 (1937).
52. S. E. Charm, J. Morningstar, C. C. Matteo, and B. Paltiel, *Anal. Biochem.*, *15*, 498 (1966).
53. H. B. Gotaas and C. G. Golueke, *Recovery of Algae from Waste Stabilization Ponds*, Algae Research Project, Sanitary Engineering Research Laboratory, Issue No 7, IER Series 44, University of California, 1957.
54. F. Schutz, *Trans. Faraday Soc.*, *38*, 85 (1942)
55. G. D. DeVivo and B. L. Karger, *Separ. Sci.*, *5*, 145 (1970).
56. D. Ellwood, *Chem. Eng.* 75, 82 (1968).
57. A. M. Gaudin, in *Froth Flotation, 50th Anniversary Volume* (D.W. Fuerstenau, ed.), AIME, New York, 1962, p. 658.
58. F. Kepak and J. Kriva, *Separ. Sci.*, *7*, 433 (1972).
59. G. Leung, Y. S. Kim, and H. Zeitlin, *Anal. Chimi. Acta*, *60*, 229 (1972).
60. Y. S. Kim and H. Zeitlin, *Chem. Communi.*, *1971*, 672. Y. S. Kim and H. Zeitlin, *Separ. Sci.*, *7*, 1 (1972). Matsuzaki and H. Zeitlin, *Ibid.*, *8*, 185 (1973).
61. E. J. Mahne and T. A. Pinfold, *J. Appl. Chem.*, *19*, 57 (1969).
62. R. B. Grieves, *J. Water Pollution Control Fed.*, *42*, R336 (1970).
63. E. J. Mahne and T. A. Pinfold, *Chem. Ind.*, *1966*, 1299.
64. A. J. Rubin and W. L. Lapp, *Separ. Sci.*, *6*, 357 (1971).
65. A. F. Taggart, *Handbook of Mineral Dressing*, Wiley, New York, 1964, p. 12: 130.
66. Ref. *10*, p. 539.
67. P. Somasundaran, *J. Colloid. Interfac Sci.*, *27*, 659 (1968).
68. P. Somasundaran and G. O. Prickett, *Trans AIME*, *244*, 369 (1969).
69. R. Eliassen and H. B. Schulhoff, *Sewage Works J.*, *16*, 287 (1944).
70. D. W. Fuerstenau, ed., *Froth Flotation, 50th Anniversary Volume*, AIME, New York, 1962.
71. I. Sheiham and T. A. Pinfold, *Separ. Sci.*, *7*, 43 (1972).

Some Modern Aspects of Ultracentrifugation

E. T. ADAMS, JR., WILL E. FERGUSON, PETER J. WAN,
JERRY L. SARQUIS, and BARNEE M. ESCOTT

CHEMISTRY DEPARTMENT
TEXAS A&M UNIVERSITY
COLLEGE STATION, TEXAS 77843

Abstract

Shortly after the ultracentrifuge was developed, it was realized that molecular-weight distributions (MWDs) of polymers could be obtained from sedimentation equilibrium experiments. Although numerous attempts have been made to obtain MWDs from sedimentation equilibrium experiments, the results were not very satisfactory, and most MWDs were obtained from sedimentation velocity experiments. Only recently have some satisfactory methods been developed for sedimentation equilibrium experiments. These methods were restricted to ideal, dilute solutions and to ultracentrifuge cells with sector-shaped centerpieces. Both of these restrictions can now be removed. Methods for correcting for nonideal behavior are shown. Procedures for obtaining MWDs from sector—or nonsector—shaped centerpieces are shown. These procedures are illustrated with real examples, and a comparison between MWDs obtained by sedimentation velocity, sedimentation equilibrium, and gel permeation chromatography experiments is shown.

Self-associations can be studied by various thermodynamic methods (osmometry, light scattering, or sedimentation equilibrium) that give average or apparent average molecular weights as a function of associating solute concentration. Of the various thermodynamic methods, the sedimentation equilibrium experiment is the best way to study self-associations. Because of the interrelation between average or apparent average molecular weights, the theory developed originally for the sedimentation equilibrium experiment can be extended to other methods. We show methods for analyzing several types of self-associations, using real examples. The advantages of thermodynamic over transport methods for studying self-associations are discussed; furthermore, we show how thermodynamic and transport experiments can be combined to yield more information about the self-associating species.

INTRODUCTION

The year 1974 marks the fiftieth anniversary of the ultracentrifuge; the first paper using the term ultracentrifuge was published by Svedberg and Rinde in 1924 (*1*). In the ensuing years the ultracentrifuge has had a wide impact in various areas of chemistry, but its most influential role has been in biochemistry, where it has been described as one of the most important research instruments in the field (*2*). According to Svedberg (*1, 3*), an ultracentrifuge has the following characteristics: it has a precise speed control; it has an optical system for viewing and/or photographing the experimental data; and it is free from convection. Another criterion encountered on modern ultracentrifuges is that they also have a good, variable range temperature control system. Instruments meeting these criteria were originally called ultracentrifuges, but are now more commonly referred to as analytical ultracentrifuges. Other high-speed centrifuges used in isolation of viruses, nucleic acids, or proteins are referred to as preparative ultracentrifuges. This article will be restricted to analytical ultracentrifugation.

Although many people believe that the ultracentrifuge is restricted to biochemistry or biophysics, this is not the case. This versatile instrument can be applied to colloid chemistry (*3*), to physical chemistry (*3–5*), to polymer chemistry (*3–5*), and to inorganic chemistry (*3–5*). It has been used to study the distribution of radii in colloidal gold sols (*1, 6*), to determine activity coefficients in silicotungstic acid solutions (*7*) and sucrose (*8*), and also to find the degree of aggregation of various salts in aqueous solutions (*9–13*). In polymer chemistry the ultracentrifuge has been used to measure sedimentation coefficients, average molecular weights, second virial coefficients, and molecular-weight distributions (*5, 15–24*). Density gradient sedimentation experiments have been used to show that the nitrogen of the deoxyribonucleic acid is divided equally between two subunits which remain intact through many generations (*25, 26*). The density gradient experiment has also been used to study greases and polymers (*27, 28*).

Because the subject of analytical ultracentrifugation is so vast, this paper will be restricted to two areas of interest to the authors: self-associations and molecular weight distributions. Both areas were considered early in the development of the ultracentrifuge (*3*), but it has only been in the last few years that real breakthroughs have been made in these areas (*16–24, 29–35*). Another reason for considering these areas is that they can be studied by other methods (*36, 37*), chromatography for example (38–41), so that comparative studies can be made. With self-associations we will show how data from two or more types of experiments can be used to extract more information about the system. For instance, values of the equilibrium constant(s), K_i, and of the monomer concentration, c_1, obtained from sedimentation equilibrium experiments can be used with sedimentation velocity experiments (performed under the same solution conditions) to evaluate the sedimentation coefficient of the associating species (42), or they might be used with analytical gel chromatography experiments to extract the partition coefficients (42). In molecular weight distribution studies, a comparison can be made with molecular weight distributions obtained from sedimentation equilibrium, sedimentation velocity, and analytical gel chromatography experiments (20). Readers interested in these and other areas of ultracentrifugation should consult the various reviews and monographs on the subject cited above.

MOLECULAR-WEIGHT DISTRIBUTIONS

Introduction

The idea of obtaining molecular-weight distributions (MWD) of non-associating, heterogeneous polymer solutions goes back to Rinde (6) in 1928. Several methods based on procedures proposed by Rinde were tried with varying success (4, 5, 24); in some cases experimental error produced negative values for the differential distribution of molecular weights, $f(M)$. Two recent developments, one by Donnelly (16, 17) and one by Scholte (18–20), have reopened interest in obtaining MWDs from sedimentation equilibrium experiments; both methods were restricted to ideal, dilute solutions and to cells with sector-shaped centerpieces. These restrictions have recently been removed (24), and it has been shown experimentally that one can obtain MWDs from nonideal solutions (43, 44). Good agreement has been observed with the MWD obtained on a dextran sample by analytical gel chromatography and by sedimentation equilibrium experiments after correction for nonideal behavior. In this section we shall describe these developments. More complete details on obtaining MWDs are to be found in papers by Adams et al. (24), by Gehatia and Wiff (23, 45–49), by Donnelly (16, 17), and by Scholte (18–20).

The Basic Sedimentation Equilibrium Equation

If one assumes that the refractive index increments, ψ_i, and the partial specific volumes, $\bar{v}_i$, for the polymeric components are the same, then one can obtain the weight-average molecular weight, M_w, or its apparent value, $M_{w\,app}$, from sedimentation equilibrium experiments. The condition of sedimentation equilibrium requires that the temperature, T, be constant and that the total potential, $\bar{\mu}_i$, of component i (3–5, 24) be constant at each radial position r, in the solution column of the ultracentrifuge cell. The quantity $\bar{\mu}_i$ is defined by

$$\bar{\mu}_i = \mu_i - \frac{M_i \omega^2 r^2}{2} \tag{1}$$

For component i, μ_i is the molar chemical potential, $-M_i\omega^2 r^2/2$ is the centrifugal potential, M_i is the molecular weight, and $\omega = 2\pi(\text{rpm})/60$ is the angular velocity of the rotor. The radial position r is restricted to distances between the position of the air-solution meniscus, r_m, and the position of the cell bottom, r_b, i.e., $r_m \leq r \leq r_b$. It is convenient to define two new quantities (5, 24)

$$\Lambda = \frac{(1 - \bar{v}\rho_0)\omega^2(r_b^2 - r_m^2)}{2RT} \tag{2}$$

and

$$\xi = \frac{r_b^2 - r^2}{r_b^2 - r_m^2} \tag{3}$$

Here $\bar{v}$ is the partial specific volume of the solute, ρ_0 is the density of the solvent, and R is the universal gas constant ($R = 8.314 \times 10^7$ ergs/deg-mole). Note that $\xi = 0$ when $r = r_b$, and $\xi = 1$ when $r = r_m$.

The sedimentation equilibrium equation for component i can be written as (15, 24)

$$-\Lambda M_i c_i = \frac{dc_i}{d\xi} - \Lambda c_i M_i \sum_k B_{ik}' c_k M_k \tag{4}$$

The quantity B_{ik}' represents a nonideal term; it is defined as

$$B_{ik}' = B_{ik} + \frac{\bar{v}}{1000 M_k} \tag{5}$$

Here c_i is the concentration (in grams/liter) of component i at radial position r, i.e., it is c_{ir}. For simplicity the subscript r will usually be dropped. It is assumed in the treatment we are using that the natural logarithm of the activity coefficient of component i can be expressed as (4, 5, 15, 24)

$$\ln y_i = M_i \sum_k B_{ik} c_k + \cdots \tag{6}$$

where

$$B_{ik} = \left(\frac{\partial \ln y_i}{\partial c_k}\right)_{T,P,c_{j\neq k}} \tag{7}$$

Most ultracentrifuges are equipped with refractometric (Rayleigh and schlieren) optics. The schlieren optical system gives information proportional to dc/dr vs r, and the Rayleigh optics give information proportional to c vs r. Figure 1 shows the type of patterns produced by the two optical systems. Since the schlieren optical system gives information proportional to dc/dr, we must sum the terms in Eq. (4) over all i solute components. Thus

$$\frac{dc}{d\xi} = \sum_i \frac{dc_i}{d\xi} = -\Lambda c M_{wr} + \Lambda \sum_i \sum_k c_i c_k B_{ik}' M_i M_k \tag{8}$$

Here

$$\frac{dc}{d\xi} = -(r_b{}^2 - r_m{}^2)\frac{dc}{d(r^2)} \tag{9}$$

and

$$M_{wr} = \sum_i c_i M_i \Big/ \sum_i c_i \tag{10}$$

Equation (8) can be rearranged to give

$$\frac{dc}{d\xi} = \frac{-\Lambda c M_{wr}}{1 + \langle B_{ik}'\rangle_r c M_{wr}} \tag{11}$$

$$= -\Lambda c M_{wr\,\text{app}}$$

provided $\langle B_{ik}'\rangle_r c M_{wr} < 1$. The quantity $\langle B_{ik}'\rangle_r$ is defined by

$$\langle B_{ik}'\rangle_r = \frac{\sum_i \sum_k c_i c_k M_i M_k B_{ik}'}{\sum_i \sum_k c_i c_k M_i M_k} \tag{12}$$

Instead of having a uniform solution as one has with light scattering or with osmometry, the centrifugal field causes a redistribution of the i polymeric components, which is why $\langle B_{ik}'\rangle_r$ varies with r. This is the term which has caused difficulty in analyzing nonideal polymer solutions by sedimentation equilibrium experiments. An ideal, dilute solution will be defined as one for which $\langle B_{ik}'\rangle_r = 0$. In this case $M_{wr\,\text{app}} = M_{wr}$ and the basic sedimentation equilibrium equation becomes

$$dc/d\xi = -\Lambda c M_{wr} \tag{13}$$

$$(\langle B_{ik}'\rangle_r = 0)$$

Rayleigh Pattern

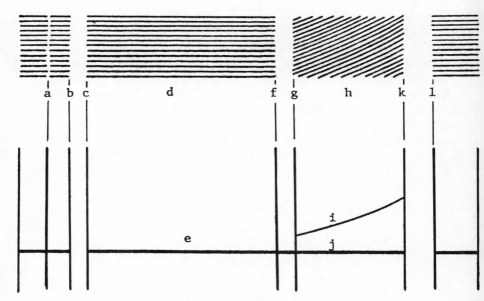

Schlieren Pattern

a – reference wire

b – inner index (outer edge)

c – top of cell

d – Rayleigh image from air column (air base line)

e – schlieren image from air column (air base line)

f – air-solvent meniscus

g – air-solution meniscus

h – Rayleigh pattern for solution

i – schlieren pattern for solution

j – solvent base line

k – bottom of cell

l – outer index (inner edge)

FIG. 1. Diagrams of the schlieren (lower) and the Rayleigh interference (upper) patterns obtained from sedimentation equilibrium experiments using double-sector centerpieces.

In order to deal with the $\langle B_{ik}' \rangle_r$ term, it is necessary to make some assumptions. Three cases are described below (43).

Case I. Assume that all the $\langle B_{ik}' \rangle$ are equal. For this case $B_{ik}' = B$ and

$$\langle B_{ik}' \rangle_r c_r M_{wr} = B c_r M_{wr} \tag{14}$$

Case II. Assume the speed effect to be small so that it can be ignored, and let $\langle B_{ik}' \rangle_r \cong B_{LS}$, the light scattering second virial coefficient. As the rotor speed goes to zero, the $c_{ir} \to c_i^0$, the initial concentration of component i. Here B_{LS} is defined by

$$B_{LS} = \frac{\sum_i \sum_k f_i f_k M_i M_k B_{ik}'}{\sum_i \sum_k f_i f_k M_i M_k} \tag{15}$$

and

$$f_i = c_i^0/c_0 \qquad (\text{or } f_k = c_k^0/c_0) \tag{16}$$

is the weight fraction of component i (or k). Note that

$$c_0 = \sum_i c_i^0 \tag{17}$$

Case III. Here a speed effect is included. We assume that

$$\langle B_{ik}' \rangle_r = B_{LS} + \left(\frac{\partial \langle B_{ik}' \rangle_r}{\partial \Lambda^2} \right)_{c_0}^0 \Lambda^2 = B_{LS} + \alpha \Lambda^2 \tag{18}$$

The superscript 0 on the partial derivative means that this quantity is evaluated in the vicinity of $\Lambda = 0$.

How does one evaluate $\langle B_{ik}' \rangle_r$? For the first two cases $\langle B_{ik}' \rangle_r$ is approximated by B_{LS}. Using Donnelly's method (16, 17), if all the B_{ik}' are equal, then there will be no speed effect on the MWD. If there is a speed effect, then there will be a difference in the MWDs determined at two or more speeds. This provides a test for the assumptions. The evaluation of B_{LS} from sedimentation equilibrium experiments has been described in detail by Wan (43) and by Adams et al. (24) for ultracentrifuge cells with sector- or nonsector-shaped centerpieces. The methods proposed by Albright and Williams (15) can be applied to either type of centerpiece. Essentially the method requires that a series of sedimentation equilibrium experiments be carried out at different speeds (three or more speeds) for each solution. Values of $M_{w \text{ cell app}}$ are calculated at each speed. These are defined by

$$M_{w \text{ cell app}} = \frac{c_b - c_m}{\Lambda c_0} \tag{19}$$

for a sector-shaped centerpiece and by

$$M_{w \text{ cell app}} = -\frac{1}{\Lambda c_0} \int_0^1 \frac{dc}{d\xi} dx \qquad (20)$$

for a nonsector-shaped centerpiece. The quantity x is the analog of ξ for a nonsector-shaped centerpiece; x is defined as (24)

$$x = \frac{\int_{r_m}^r f(r) \, dr}{\int_{r_m}^{r_b} f(r) \, dr} \qquad (21)$$

Here $f(r)$ is a function that describes how the cross-sectional width of the cell varies with radial distance r. Figure 2 shows the top views of a double-sector centerpiece and of a multichannel equilibrium centerpiece. Once one has run the experiments at different speeds it is necessary to evaluate

$$M_w^0 {}_{\text{cell app}} = \lim_{\Lambda \to 0} M_{w \text{ cell app}} = M_w^0 {}_{\text{app}}$$

This can be done by plotting $M_{w \text{ cell app}}$ vs Λ and extrapolating the plot to $\Lambda = 0$; such a plot is shown in Fig. 3. The intercept of this plot gives $M_w^0 {}_{\text{cell app}}$. Alternatively one can plot $\Delta c/c_0$ vs Λ or $-\int_0^1 (dc/d\xi) \, dx/c_0$ vs Λ depending on the type centerpiece one uses (24). The limiting slope of these plots as shown in Fig. 4, gives $M_w^0 {}_{\text{app}}$ for each c_0. Note that these plots must go through the origin; for example, with a sector-shaped centerpiece $\Delta c = 0$ when $\Lambda = 0$. Note also that at constant c_0, $\Delta c/c_0$ is a function of Λ. Thus one could use regression analysis to obtain a polynomial of the form

$$\Delta c/c_0 = \alpha + \beta \Lambda + \gamma \Lambda^2 + \cdots \qquad (22)$$

The quantity α should be equal to zero and β can be identified with $M_w^0 {}_{\text{app}}$.

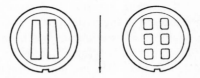

FIG. 2. Top view of double-sector and multichannel equilibrium centerpieces. One side of either centerpiece is reserved for the solvent or buffer solution. The arrow indicates the direction of increasing centrifugal field strength. With multichannel centerpieces the most dilute solution is put in one of the centrifugal or outermost holes, and the most concentrated solution is put in one of the centripetal or innermost holes.

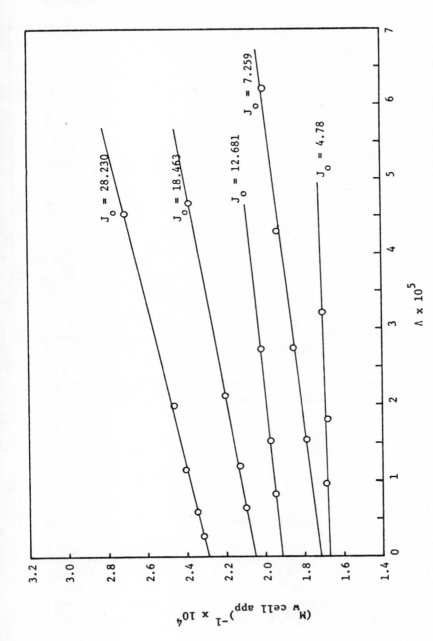

FIG. 3. Plot of $1/M_{w\text{ cell app}}$ vs Λ for a dextran T-70 sample dissolved in water ($T = 25°C$). The initial concentrations, J_0, are given in fringes (12 mm centerpiece, $\lambda = 546$ nm); at $25°C$, $J = 3.297c$ for c in g/l. The intercept of each plot gives $1/M^0_{w\text{ app}}$ for each J_0.

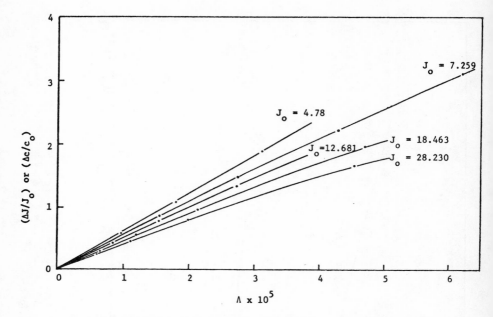

FIG. 4. Plots of $\Delta J/J_0$ vs Λ. The limiting slope of each plot gives $M_{w\ app}^0$. Dextran T-70 dissolved in water. $T = 25\,°C$.

Fujita (5) has shown that the sedimentation equilibrium second virial coefficient, $\langle B_{ik}' \rangle$, depends on Λ^2. With the use of Eq. (18) one notes that at constant c_0,

$$\lim_{\Lambda \to 0} 1/M_{w\ cell\ app} = \lim_{\Lambda \to 0}\left[\frac{1}{M_w} + (B_{LS} + \alpha\Lambda^2)c_0 + \cdots\right]$$
$$= 1/M_w + B_{LS}c_0 = 1/M_{w\ app}^0 \tag{23}$$

and that

$$\lim_{\Lambda \to 0} \frac{\partial}{\partial \Lambda^2}\bigg|_{c_0} (1/M_{w\ cell\ app}) = \alpha c_0 \tag{23a}$$

Equation (23a) shows how one might try to evaluate α. The quantity B_{LS} is obtained from the slope of a plot of $1/M_{w\ app}^0$ vs c_0 (see Eq. 23); a plot of this kind is shown in Fig. 5. Thus the required quantities for the non-ideal correction can be evaluated from the experimental data. Having an estimated value of the $\langle B_{ik}' \rangle_r$, it is possible to calculate ideal values of c

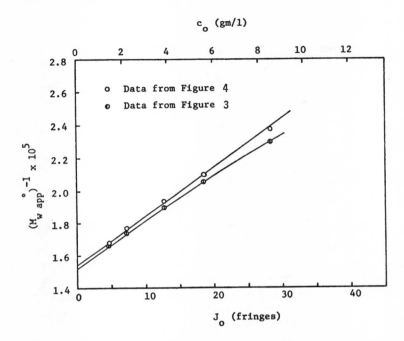

FIG. 5. Plots of $1/M_w^0{}_{app}$ vs J_0 for the dextran T-70 sample using data from Figs. 3 and 4. The intercept of this plot gives $1/M_w$. The average value of M_w is 6.59×10^4 Daltons.

or dc/dr (24). Equation (11) can be rearranged to give

$$\frac{1}{cM_{wr}} \equiv \frac{-1}{(dc/d\xi)_{\text{ideal}}} = \frac{1}{cM_{wr\ app}} - \frac{\langle B_{ik}' \rangle_r}{\Lambda} \tag{24}$$

The values of $(dc/d\xi)_{\text{ideal}}$ so obtained can be used with Scholte's method or any of the other methods for obtaining MWDs. For Donnelly's method the quantity $(d \ln c/d\xi)_{\text{ideal}}$ is needed. Thus Eq. (11) can be rearranged to give

$$\frac{1}{M_{wr}} \equiv \frac{-1}{\left[\dfrac{1}{\Lambda} \dfrac{d \ln c}{d\xi} \right]_{\text{ideal}}}$$

$$= \frac{1}{M_{wr\ app}} - \langle B_{ik}' \rangle_r c \tag{24a}$$

Here

$$-\frac{1}{\Lambda}\frac{d\ln c}{d\xi} = M_{wr\ app} \tag{25}$$

An additional quantity required for Donnelly's method is $c_0/[c(\xi = 1)]_{ideal}$. This is readily obtained from

$$\frac{c_0}{c(\xi = 1)_{ideal}} = \int_0^1 \left[\frac{c(\xi)}{c(\xi = 1)}\right]_{ideal} d\xi \tag{26}$$

Note that

$$\int_1^\xi \left(\frac{d\ln c}{d\xi}\right)_{ideal} d\xi = \ln\left[\frac{c(\xi)}{c(\xi = 1)}\right]_{ideal} \tag{27}$$

Concentration and Concentration Gradient Distributions

One of the unique features of the sedimentation equilibrium experiment is that it allows the evaluation of average molecular weights without any prior knowledge of the MWD. In addition it is also possible to obtain the MWD from sedimentation equilibrium experiments. No other physical method for studying macromolecules has this versatility. We have shown in the previous section how nonideal effects can be accounted for, so it will be assumed that nonideal corrections have been applied, and in the discussion that follows equations applicable to ideal dilute solution conditions will be used.

For ideal, dilute solution conditions, Eq. (4) becomes (*4, 5, 24*)

$$\frac{d\ln c_i}{d\xi} = -\Lambda M_i \tag{28}$$

This equation can be integrated between $\xi = 1$ and $\xi = \xi$ and then recast in exponential form to give

$$c_i(\xi) = c_i(\xi = 1)\exp\left[-\Lambda M_i \xi\right] \tag{29}$$

From Eq. (29) one notes that the concentration distribution for component i is in exponential form. Different values of M_i will give different concentration distributions and also different concentration gradient distributions; thus one will expect different values for any average molecular weight at any radial position. The weight average molecular weight at any radial position, M_{wr}, is given by Eq. (13) or its more familiar form (*3–5, 24, 50*)

$$\frac{d\ln c}{d(r^2)} = AM_{wr} \tag{30}$$

where

$$A = (1 - \bar{v}\rho)\omega^2/2RT \tag{31}$$

The z-average molecular weight at any radial position, M_{zr}, is obtained from $(3, 50)$

$$M_{zr} = \frac{d(cM_{wr})}{dc} = M_{wr} + c\frac{dM_{wr}}{dc} \tag{32}$$

Here

$$M_{zr} = \frac{\sum_i c_{ir}M_i^2}{\sum_i c_{ir}M_i} \tag{33}$$

In principle it is possible to obtain any higher average molecular weight, since $(3–5, 24, 50)$

$$M_{(z+q)r} = \frac{d(cM_wM_zM_{z+1} \cdots M_{(z+q-1)})_r}{d(cM_wM_zM_{z+1} \cdots M_{(z+q-2)})_r} \tag{34}$$

Note that

$$M_{(z+q)r} = \frac{\sum c_{ir}M_i^{(q+2)}}{\sum c_{ir}M_i^{(q+1)}} \qquad q = 0, 1, 2, \ldots \tag{35}$$

In practice it is virtually impossible to go beyond M_{zr}, since numerical differentiation is involved. The number-average molecular weight at any radial position, M_{nr}, is not readily obtainable, since $(50, 51)$

$$\frac{c_r}{M_{nr}} - \frac{c_{rm}}{M_{nr_m}} = A \int_{r_m}^{r} cd(r^2) \tag{36}$$

Here two unknowns, M_{nr} and M_{nr_m}, are involved. Even though this equation can be transformed to one equation in one unknown, it is still difficult to obtain M_{nr_m} (50).

The quantities M_{wr} and M_{zr} are useful in the analysis of self-association; they can also be used to develop methods for the evaluation of the cell averages $M_{w\,cell}$ and $M_{z\,cell}$. For nonassociating polymers in ideal, dilute solutions, one can obtain the M_w and M_z of the polymer from measurements of $M_{w\,cell}$ and $M_{z\,cell}$. In order to evaluate $M_{w\,cell}$ or $M_{z\,cell}$, it is necessary to know the shape of the cell (centerpiece), since the conservation of mass equations enter into these equations. The conservation of mass states that the total amount of solute in a closed system is constant at any time. The mass of solute is given by $(4, 5, 24, 50)$

$$\text{mass solute} = \int_{r_m}^{r_b} c_r \, dV \tag{37}$$

Sector-Shaped Centerpieces

For a sector-shaped centerpiece

$$dV = \frac{\theta h}{2} d(r^2) \tag{38}$$

where θ is the sector angle and h is the cell thickness. The amount of solute at the beginning of the experiment is given by

$$\text{mass solute} = \frac{\theta h}{2} c_0(r_b^2 - r_m^2) \tag{39}$$

and at sedimentation equilibrium it is given by

$$\text{mass solute} = \frac{\theta h}{2} \int_{r_m}^{r_b} c_r \, d(r^2) \tag{40}$$

Setting Eqs. (39) and (40) equal to each other, one obtains (3, 5, 24)

$$c_0(r_b^2 - r_m^2) = \int_{r_m}^{r_b} c_r \, d(r^2) \tag{41}$$

or

$$c_0 = \int_0^1 c(\xi) \, d\xi \tag{41a}$$

In order to evaluate c_m, the meniscus concentration, one must note that

$$(c_0 - c_m)(r_b^2 - r_m^2) = \int_{r_m}^{r_b} (c_r - c_m) \, d(r^2) \tag{42}$$

The quantity $c_r - c_m$ is the difference in concentration between the concentration at r and that at r_m; it is directly proportional to the difference in Rayleigh interference fringes between these two radial positions. The quantity $M_{w \, \text{cell}}$ is defined by (3–5, 24, 51)

$$M_{w \, \text{cell}} = \frac{\int_{r_m}^{r_b} M_{wr} c \, d(r^2)}{\int_{r_m}^{r_b} c_r \, d(r^2)} \tag{43}$$

Since $cM_{wr} = \sum_i c_{ir} M_i$, one notes that

$$\begin{aligned}
M_{w \, \text{cell}} &= \frac{\sum_i M_i \int_{r_m}^{r_b} c_i \, d(r^2)}{c_0(r_b^2 - r_m^2)} \\
&= \frac{\sum_i M_i c_i^0 (r_b^2 - r_m^2)}{c_0(r_b^2 - r_m^2)} = \frac{\sum_i c_i^0 M_i}{\sum_i c_i^0} \\
&= M_w \text{ of the original solution}
\end{aligned} \tag{44}$$

The insertion of Eq. (30) into the numerator of Eq. (43) leads to

$$
M_{w \, \text{cell}} = \frac{\int_{r_m}^{r_b} \dfrac{d \ln c}{d(r^2)} c \, d(r^2)}{A c_0 (r_b^2 - r_m^2)} = \frac{c_b - c_m}{\Lambda c_0}
$$

$$
[\Lambda = A(r_b^2 - r_m^2)] \tag{45}
$$

The quantity $M_{z \, \text{cell}}$ is defined by (3–5, 24, 51)

$$
M_{z \, \text{cell}} = \frac{\int_{r_m}^{r_b} M_{zr} \, dc}{\int_{c_m}^{c_b} dc} = \frac{\left[\left(\dfrac{1}{r} \dfrac{dc}{dr} \right)_{r_b} - \left(\dfrac{1}{r} \dfrac{dc}{dr} \right)_{r_m} \right]}{2A(c_b - c_m)} \tag{46}
$$

Here M_{zr} is given by (20)

$$
M_{zr} = \frac{1}{2A} \frac{d \left(\dfrac{1}{r} \dfrac{dc}{dr} \right)}{dc} \tag{47}
$$

One can use arguments similar to those used in Eq. (44) to show that $M_{z \, \text{cell}}$ is the M_z of the polymer. For nonideal solutions one obtains $M_{w \, \text{app}}$ and $M_{z \, \text{app}}$ from Eqs. (45) and (46). It can be shown that as $\Lambda \to 0$, $M_{w \, \text{app}} \to M_{w \, \text{app}}^0$, where

$$
M_{w \, \text{app}}^0 = \frac{M_{w \, \text{cell}}}{1 + B_{LS} c_0 M_{w \, \text{cell}}} \tag{48}
$$

A similar relation can be developed for $M_{z \, \text{app}}$ (24, 43).

Nonsector-Shaped Centerpieces

With nonsector-shaped centerpieces dV is given by

$$
dV = hf(r) \, dr \tag{49}
$$

and the conservation of mass equation (see Eq. 37) becomes (3, 24)

$$
c_0 \int_{r_m}^{r_b} f(r) \, dr = \int_{r_m}^{r_b} c_r f(r) \, dr \tag{50}
$$

or using dx (see Eq. 21)

$$
c_0 = \int_0^1 c(\xi) \, dx \tag{51}
$$

To evaluate c_m it is noted that

$$
(c_0 - c_m) \int_{r_m}^{r_b} f(r) \, dr = \int_{r_m}^{r_b} (c_r - c_m) f(r) \, dr \tag{52}
$$

or using dx

$$(c_0 - c_m) = \int_0^1 [c(\xi) - c(\xi = 1)] \, dx \tag{53}$$

The quantity $M_{w\,cell}$ is defined by (4, 18)

$$M_{w\,cell} = \frac{\int_{r_m}^{r_b} M_{wr} c_r f(r) \, dr}{\int_{r_m}^{r_b} c_r f(r) \, dr}$$

$$= \frac{\dfrac{1}{2A} \displaystyle\int_{r_m}^{r_b} \dfrac{f(r) \, dc}{r}}{c_0 \displaystyle\int_{r_m}^{r_b} f(r) \, dr} = \frac{-1}{\Lambda c_0} \int_0^1 \frac{dc}{d\xi} \, dx \tag{54}$$

If the solution is nonideal, then Eq. (54) gives $M_{w\,app}$. The limit of $M_{w\,app}$ as $\Lambda \to 0$ becomes $M_{w\,app}^0$, which is defined by Eq. (48). The quantity $M_{z\,cell}$ is defined by (13, 50)

$$M_{z\,cell} = \frac{\displaystyle\int_{r_m}^{r_b} M_{zr} f(r) \frac{dc}{r}}{\displaystyle\int_{r_m}^{r_b} f(r) \frac{dc}{r}}$$

$$= \frac{\dfrac{1}{2A} \displaystyle\int_{r_m}^{r_b} \dfrac{f(r)}{r} \left[d\left(\dfrac{1}{r} \dfrac{dc}{dr} \right) \middle/ dc \right] dc}{\displaystyle\int_{r_m}^{r_b} \dfrac{f(r)}{r} \, dc} \tag{55}$$

Donnelly's Method for MWDs (16, 17)

This method is based on data obtained from a single sedimentation equilibrium experiment. It is quite good with unimodal MWDs, but it may not be as good as Scholte's method with multimodal MWDs. The starting equation here is Eq. (29). In order to make it more useful, it is necessary to relate c_i ($\xi = 1$) to the initial concentration of i, c_i^0; this requires the application of the conservation of mass equation. Thus this method must be considered separately for each of the two kinds of center-pieces.

Sector-Shaped Centerpieces

For component i the conservation of mass equation becomes

$$c_i^0 = \int_0^1 c_i(\xi) \, d\xi \tag{56}$$

The insertion of Eq. (29) in Eq. (56) and subsequent rearrangement leads to

$$c_i(\xi = 1) = \frac{\Lambda M_i c_i^{\,0}}{1 - \exp(-\Lambda M_i)} \tag{57}$$

The substitution of Eq. (57) into Eq. (29) yields

$$c_i(\xi) = \frac{\Lambda M_i c_i^{\,0} \exp(-\Lambda M_i \xi)}{1 - \exp(-\Lambda M_i)} \tag{58}$$

Summation over the i solute components followed by division of both sides by c_0 leads to (16, 17, 24, 50)

$$\theta(\xi) = \frac{c(\xi)}{c_0} = \sum_i \frac{\Lambda M_i f_i \exp(-\Lambda M_i \xi)}{1 - \exp(-\Lambda M_i)} \tag{59}$$

Here $f_i = c_i^{\,0}/c_0$ is the weight fraction of component i. Equation (59) can be differentiated with respect to ξ to give

$$-\theta'(\xi) = \frac{-1}{c_0} \frac{dc(\xi)}{d\xi} = \sum_i \frac{\Lambda^2 M_i^{\,2} f_i \exp(-\Lambda M_i \xi)}{1 - \exp(-\Lambda M_i)}$$
$$= U(\Lambda, \xi) \tag{60}$$

Equations (59) and (60) are used in Scholte's (18–20) method. Now if one assumes that a continuous distribution of molecular weights is present, then f_i is replaced by $f(M)\,dM$ which is the weight fraction of solute having molecular weights between M and $M + dM$, and the summation is replaced by an integral running between 0 and ∞. Thus Eqs. (59) and (60) become (16, 17, 24, 50)

$$\theta(\xi) = \int_0^\infty \frac{\Lambda M f(M) \exp(-\Lambda M \xi)\,dM}{1 - \exp(-\Lambda M)} \tag{61}$$

and

$$\theta'(\xi) = -\frac{1}{c_0} \frac{dc(\xi)}{d\xi} = \int_0^\infty \frac{\Lambda^2 M^2 f(M) \exp(-\Lambda M \xi)\,dM}{1 - \exp(-\Lambda M)} \tag{62}$$

Here $f(M)$ is the differential distribution of molecular weights; this is the quantity we want to determine. Donnelly pointed out that Eqs. (61) and (62) are Laplace transforms (16, 17); if one can find an analytical expression for the Laplace transform, then the MWD can be obtained from the inverse of the Laplace transform. In order to see this more clearly, in Eq. (16) let $\xi = s$, $t = \Lambda M$, and

$$\phi(t) = \frac{\Lambda M f(M)}{1 - \exp(-\Lambda M)} = \frac{t f(M)}{1 - \exp(-t)} \tag{63}$$

Then it follows that

$$L\{\phi(t)\} = \Lambda\theta = \int_0^\infty \phi(t)e^{-st}\, dt = f(s) \tag{64}$$

Here $L\{\phi(t)\}$ is the symbol for the Laplace transform of the function $\phi(t)$. In order to use $L\{\phi(t)\}$, we must find an analytical expression for it; in other words, what is the form of $f(s)$? Donnelly solved this neatly (*16, 17, 24*). Let us define

$$F(n, u) = \frac{1}{\left(\dfrac{d\ln c}{d(r^2)}\right)} \tag{65}$$

where

$$u = \frac{r^2 - r_m{}^2}{r_b{}^2 - r_m{}^2} = 1 - \xi = 1 - s \tag{66}$$

Suppose that one makes a plot of $F(n, u)$ vs u, and suppose that this plot is a straight line of the form

$$F(n, u) = P - Qu \tag{67}$$

Here P is the intercept at $u = 0$ and Q is the slope. Now note that

$$du = \frac{d(r^2)}{r_b{}^2 - r_m{}^2} \tag{68}$$

or

$$d(r^2) = (r_b{}^2 - r_m{}^2)\, du = b\, du \tag{68a}$$

The integral

$$\int_{r_m}^r \frac{d\ln c}{d(r^2)}\, d(r^2) = \ln\frac{c_r}{c_{r_m}} = \ln\frac{c(\xi)}{c(\xi = 1)} \tag{69}$$

and

$$\exp\int_{r_m}^r \frac{d\ln c}{d(r^2)}\, d(r^2) = \frac{c(\xi)}{c(\xi = 1)} \tag{70}$$

If Eq. (70) is multiplied by $\Lambda c(\xi = 1)/c_0 = 1/K$, then we obtain $\Lambda\theta(\xi)$. Thus

$$\begin{aligned}
L\{\phi(t)\} = \Lambda\theta(\xi) &= \frac{\Lambda c(\xi = 1)}{c_0}\exp\int_{r_m}^r \frac{d\ln c}{d(r^2)}\, d(r^2) \\
&= \frac{1}{K}\exp\int_{r_m}^r \frac{d\ln c}{d(r^2)}\, d(r^2) = \frac{1}{K}\exp\int_0^u \frac{b\, du}{F(n, u)}
\end{aligned} \tag{71}$$

The insertion of Eqs. (65) and (68a) into Eq. (69) leads to

$$\int_{r_m}^{r} \frac{d \ln c}{d(r^2)} d(r^2) = \int_{0}^{u} \frac{b \, du}{F(n, u)} \tag{72}$$

Since $F(n, u) = P - Qu$, one notes that

$$\int_{0}^{u} \frac{b \, du}{P - Qu} = \frac{b}{Q} \ln \frac{P}{P - Qu} \tag{73}$$

Using Eq. (66) this becomes

$$\frac{b}{Q} \ln \frac{P}{P - Qu} = \frac{b}{Q} \ln \frac{P}{Q[(P/Q) - 1 + s]} \tag{73a}$$

Therefore

$$L\{\phi(t)\} = \Lambda\theta = \frac{1}{K} \left[\frac{P}{Q\{(P/Q) - 1 + s\}} \right]^{b/Q}$$
$$= \frac{(P/Q)^{b/Q}}{K} \left[\frac{1}{s + a} \right]^{n} \tag{74}$$

where $a = (P/Q) - 1$ and $n = b/Q$.
A well-known mathematical relation states that (16, 17, 24)

$$\frac{\Gamma(n)}{(s + a)^n} = \int_{0}^{\infty} t^{n-1} e^{-(s+a)t} \, dt \tag{75}$$

where $\Gamma(n)$ is the gamma function of n. Thus it follows that

$$L\{\phi(t)\} = \int_{0}^{\infty} \phi(t) e^{-st} \, dt$$
$$= \int_{0}^{\infty} \frac{(P/Q)^{b/Q}}{K} \frac{t^{(b/Q)-1}}{\Gamma(b/Q)} \exp\{-[(P/Q) - 1]t - st\} \, dt \tag{76}$$

Comparison of the two integrals in Eqs. (75) and (76) indicates that

$$\phi(t) = \frac{(P/Q)^{b/Q}}{K} \frac{t^{(b/Q)-1}}{\Gamma(b/Q)} e^{-[(P/Q)-1]t} \tag{77}$$

It follows from Eq. (63) that

$$f(M) = \frac{(P/Q)^{b/Q}}{K} \frac{t^{(b/Q)-2}}{\Gamma(b/Q)} e^{-[(P/Q)-1]t}(1 - e^{-t}) \tag{78}$$

Nonsector-Shaped Centerpieces (24, 43)

Here the conservation of mass equation for component i is (24, 43)

$$c_i^0 = \int_0^1 c_i \, dx \tag{79}$$

The insertion of Eq. (29) into Eq. (79) leads to

$$c_i(\xi = 1) = \frac{c_i^0 \exp(-\Lambda M_i \xi)}{\int_0^1 [\exp(-\Lambda M_i \xi)] \, dx} \tag{80}$$

The analogs of Eqs. (59) and (60) become (24, 43)

$$\hat{\theta}(\xi) = \frac{c(\xi)}{c_0} = \sum_i \frac{f_i \exp(-\Lambda M_i \xi)}{\int_0^1 [\exp(-\Lambda M_i \xi)] \, dx} \tag{81}$$

and

$$-\hat{\theta}'(\xi) = \frac{-1}{c_0} \frac{dc(\xi)}{d\xi} = \sum_i \frac{\Lambda M_i f_i \exp(-\Lambda M_i \xi)}{\int_0^1 [\exp(-\Lambda M_i \xi)] \, dx}$$
$$= V(\Lambda, \xi) \tag{82}$$

For a continuous distribution of molecular weights, $\hat{\theta}(\xi)$ becomes

$$\hat{\theta}(\xi) = \int_0^\infty \frac{f(M) \exp(-\Lambda M \xi) \, dM}{\int_0^1 [\exp(-\Lambda M \xi)] \, dx} \tag{83}$$

Letting $t = \Lambda M$, $\xi = s$, and

$$\gamma(t) = \frac{f(M)}{\int_0^1 [\exp(-\Lambda M \xi)] \, dx} \tag{84}$$

one can show that the Laplace transform, $L\{\gamma(t)\}$, of $\gamma(t)$ is given by (24, 43)

$$\Lambda \hat{\theta}(\xi) = L\{\gamma(t)\} = \int_0^\infty \gamma(t) e^{-st} \, dt \tag{85}$$

Now we can follow the previous procedure. $F(n, u)$ (see Eq. 65) is defined in the same manner, and Eq. (71) applies to this case also. So, if $F(n, u) = P - Qu$, then $L\{\gamma(t)\}$ will also be defined by Eq. (74). However, $f(M)$ in this case becomes

$$f(M) = \frac{(P/Q)^{b/Q}}{K} \frac{t^{(b/Q)-1}}{\Gamma(b/Q)} e^{-[(P/Q)-1]t} \int_0^1 [\exp(-\Lambda M \xi)] \, dx \tag{86}$$

Thus the only difference in the two methods is that

$$f(M) = \left[\frac{1 - \exp(-\Lambda M)}{\Lambda M}\right]\phi(t) \tag{87}$$

for a sector-shaped centerpiece, and

$$f(M) = \left\{\int_0^1 \exp(-\Lambda M\xi)\, dx\right\}\gamma(t) \tag{88}$$

for a nonsector-shaped centerpiece. Table I in the paper by Adams et al. (24) gives the Laplace transform for three different equations for $F(n, u)$; a fourth case is described by Donnelly. For other situations not covered by these four cases, one might be able to use the complex inversion method to find the inverse of the Laplace transform. Although the mathematical treatment may look formidable, this is really a beautiful and simple method to use.

Results with Donnelly's Method

Figure 6 shows plots of $F(n, u)$ vs u (see Eq. 65) for a dextran sample dissolved in water. The sedimentation equilibrium experiments were run at 25°C and at 8000 rpm. The upper plot has not been corrected for

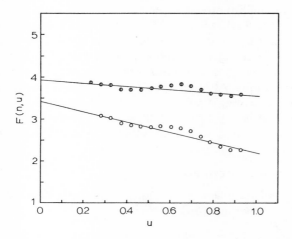

FIG. 6. Plot of $F(n, u)$ vs u for Dextran T-70 in water. $T = 25\,°C$. This plot is required for Donnelly's method for obtaining MWDs. These experiments were carried out in multichannel, equilibrium centerpieces. The upper plot ($\odot$) uses data uncorrected for nonideal behavior, whereas the lower plot ($\bigcirc$) uses data corrected for nonideal behavior assuming $B_{LS} = \langle B_{lk}'\rangle_r$.

nonideality, whereas the lower plot has been corrected for nonideality. We then assumed $\langle B_{ik}' \rangle_r \equiv B_{LS}$, where B_{LS} was calculated from experiments at different speeds. Equation (20) was used to calculate $M_{w \text{ cell app}}$ at each speed. These results were extrapolated to zero speed, and B_{LS} was determined from a plot of $1/M_{w \text{ app}}^0$ vs c_0 (see Eq. 24). The value of B_{LS} used to obtain $(d \ln c/d\xi)_{\text{ideal}}$ (see Eq. 24a) was 0.322×10^{-6} for concentrations in terms of green (12 mm) fringes. With the aid of Eqs. (26) and (27), one could proceed with the analysis described in the preceding section for nonsector-shaped cells. Figure 7 shows the MWD obtained by Donnelly's method with sector- and nonsector-shaped cells. Note that the corrected values agree with the manufacturer's MWD which was obtained by a combination of analytical gel chromatography and light scattering. Also note that the uncorrected MWDs in no way resemble the corrected or manufacturer's MWD. Figure 6 shows the difference in corrected and uncorrected plots of $F(n, u)$ vs u; the correction makes quite a difference in the MWD.

Scholte's Method (18–20)

This method is based on Eq. (60) for a sector-shaped centerpiece or Eq. (82) for a nonsector-shaped centerpiece and requires that one perform sedimentation equilibrium experiments at different speeds on the same solution. After the Rayleigh and schlieren data have been recorded at one speed, the speed is changed, and the solution is allowed to come to sedimentation equilibrium again; the photographic data are collected at each new speed before going on to the next speed. This method uses the field to fractionate the sample. At lower speeds the concentration distribution of the lower molecular weight solutes is relatively smaller than that for the high molecular weight solutes. At higher speeds the larger molecular weight solutes are pushed toward the cell bottom and the lower molecular weight solutes are distributed throughout the cell.

Sector-Shaped Centerpieces

The starting equation here is Eq. (60). Scholte designates the derivative as $U(\Lambda, \xi)$; thus he writes (18–20, 24)

$$
\begin{aligned}
U(\Lambda, \xi)_j &= -\frac{1}{c_0} \frac{dc(\xi)}{d\xi} \\
&= \sum_i f_i \frac{\Lambda_j^2 M_i^2 \exp(-\Lambda_j M_i \xi)}{1 - \exp(-\Lambda_j M_i)} + \delta_j \\
&= \sum_i f_i K_{ij} + \delta_j
\end{aligned}
\tag{89}
$$

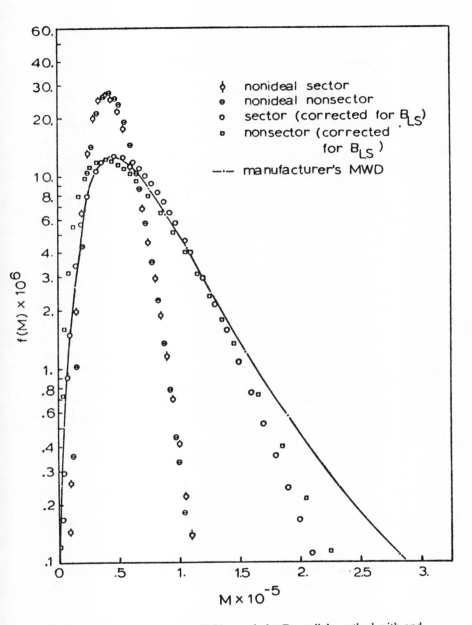

FIG. 7. MWD for the Dextran T-70 sample by Donnelly's method with and without the B_{LS} correction for nonideality. Note how much better the corrected data agrees with the manufacturer's MWD (solid line), which was obtained by analytical gel chromatography. An even better agreement is obtained when a rotor speed correction is included. The data in these plots were collected using sector- and nonsector-shaped centerpieces.

where

$$K_{ij} = \frac{\Lambda_j{}^2 M_i{}^2 \exp(-\Lambda_j M_i \xi)}{1 - \exp(-\Lambda_j M_i)} \tag{90}$$

and δ_j is a term expressing experimental error. The quantities subscripted by i depend on the molecular weights, M_i, that are chosen; thus f_i will be the weight fraction of component i whose molecular weight is M_i. The quantity Λ_j has the subscript j to indicate that this quantity depends on the speed used and not on the molecular weight of component i. The experimental data are usually read at five evenly spaced values of ξ, i.e., at $\xi = 0$, 1/4, 1/2, 3/4, and 1, although at some speeds one will not be able to read data at all of the values of ξ. In order to use Scholte's method (18–20), one must assume a discrete range of molecular weights, M_i, which will bracket the sample. For example, the molecular weights chosen for this first range or first series might be $M_1 = 2,500$, $M_2 = 5,000$, $M_3 = 10,000$, and so on until a molecular weight larger than the largest expected value for the polymer is attained. It was found by experience that the interval between molecular weights is logarithmic, and that the series is best constructed so that each successive molecular weight chosen has twice the value of the molecular weight immediately preceeding it. The speeds chosen should be such that the ratio of successive ω^2 is two, or as close to this ratio as possible. The values of $U(\Lambda, \xi)$ that are available should exceed the number of M_i that are chosen. The first choice of molecular weights is called the first series. Once the first series is chosen, the $U(\Lambda, \xi)$ values can be used to write down a set of linear equations.

$$\begin{aligned}
U(\Lambda, \xi)_1 &= f_1 K_{11} + f_2 K_{21} + f_3 K_{31} + \cdots + \delta_1 \\
U(\Lambda, \xi)_2 &= f_1 K_{12} + f_2 K_{22} + f_3 K_{32} + \cdots + \delta_2 \\
&\;\;\vdots \qquad\quad \vdots \qquad\quad \vdots \qquad\quad \vdots \qquad\qquad \vdots \\
U(\Lambda, \xi)_n &= f_1 K_{1n} + f_2 K_{2n} + f_3 K_{3n} + \cdots + \delta_n
\end{aligned} \tag{91}$$

Scholte (18–20) solves this set of equations by a linear programming technique. These equations are perfectly adapted to linear programming, since (1) we are dealing with a set of linear equations involving only positive quantities; (2) $\sum_i f_i = 1$, i.e., the sum of all the weight fractions must be one; (3) each f_i must satisfy the condition $0 \le f_i \le 1$; and (4) the f_i's are chosen so the sum of the absolute value of the error is a minimum, i.e., $\sum_j |\delta_j|$ is a minimum. The f_i are determined on a computer.

The set of f_i obtained from the computer program for the first series of molecular weights is not unique. One could choose another set of M_i's, and this is just what Scholte does. A second series of M_i is obtained by multiplying each M_i in the first series by $2^{1/4}$, i.e., M_i(2nd series) $= 2^{1/4} \times M_i$(1st series). Another array of $U(\Lambda, \xi)_j$ is set up for the second series, and again the computer is used to obtain the f_i using the linear programming technique. This procedure is repeated with a third series, M_i(3rd series) $= 2^{1/2} \times M_i$(1st series), and a fourth series, M_i (4th series) $= 2^{3/4} \times M_i$(1st series). Table 1 shows an example of the type of data used and the results obtained from the computer program for the first and third series. This data is taken from the work of Scholte (*18*). The material used was a polyethylene sample which was dissolved in biphenyl; the experiments were performed at the theta temperature, 123.2°C. Figure 8 shows the MWD that was obtained; note that Scholte plots his MWD as $Mf(M)$ vs M. It is quite evident from this figure that Scholte's method can detect a multimodal MWD.

How does one obtain the MWD? First of all note that

$$\sum_i f_i \text{ (any series)} = 1 \tag{92}$$

$$\sum_i f_i \text{ (all four series)} = 4 \tag{92a}$$

and

$$\sum_i f_i/4 \text{ (for all four series)} = 1 \tag{92b}$$

Thus Eq. (92b) is also a solution, and it could be used in obtaining the plot of $f(M)$ vs M since more points would be available. In order to obtain a plot of $f(M)$ vs M note that

$$\sum_i f_i \text{ (any series)} = 1 = \int_0^\infty f(M)\, dM \tag{93}$$

TABLE 1a
Tabulation of Raw Data Needed for Molecular Weight Distribution

$\Lambda \times 10^6$	ξ^a				
	0	1/4	1/2	3/4	1
2.5	0.294	0.260	0.232	0.208	0.187
10	2.121	1.237	0.799	0.554	0.405
40			1.346	0.639	0.372
160			0.959	0.337	0.144

^{a}Note that $\Lambda = (1 - \bar{v}\rho)\omega^2(r_b^2 - r_m^2)/2RT$.

$$\xi = \frac{r_b^2 - r^2}{r_b^2 - r_m^2}$$

$\xi = 0$ when $r = r_b$.
$\xi = 1$ when $r = r_m$.

TABLE 1b
Results Obtained from Computer Program[a]

First series			Third series		
M_i	f_i	$M_i^2 f(M)_i$	M_i	f_i	$M_i^2 f(M)_i$
8,839	0	0	12,450	0	0
17,678	0.031	790.8	24,900	0.075	2,694.81
35,356	0.084	4,285.6	49,800	0.149	10,707.36
70,712	0.313	31,937.7	99,600	0.453	65,106.49
141,424	0.450	91,833.8	199,200	0.291	83,646.75
282,848	0.104	42,447.6	398,400	0.011	6,323.81
565,696	0	0	796,800	0.012	13,797.40
1,131,392	0.015	24,489.0	1,593,600	0.009	20,696.10
2,262,784	0.0021	6,856.9	3,187,200	0.0002	919.83
Σf_i	0.9991		Σf_i	1.0002	

[a]For the second series (not tabulated): M_i(2nd series) = M_i(1st series) $\times 2^{1/4}$. For the third series: M_i(3rd series) = M_i(1st series) $\times 2^{1/2}$. For the fourth series: M_i(4th series) = M_i(1st series) $\times 2^{3/4}$.

Note that

1. Σf_i(1st series) = 0.9991.
2. Σf_i(2nd series) = 1.002.
3. $\displaystyle \int_0^\infty M f(M) \frac{dM}{M} = 1 \simeq \Delta \ln M \, \Sigma [M f(M)]_i$

$$= \frac{0.693}{4} \sum_{\text{all 4 series}} \frac{f_i}{0.693}$$

4. $\displaystyle M_w = \int_0^\infty M f(M) \, dM = \int_0^\infty M^2 f(M) \frac{dM}{M}$

$$\simeq \Delta \ln M \sum_{\text{all 4 series}} [M^2 f(M)]_i$$

$$= \frac{0.693}{4} \sum_{\text{all 4 series}} \frac{M_i f_i}{0.693}$$

Using data from the first and third series

$$\Delta \ln M = \frac{0.693}{2} = 0.346 = \frac{1}{2} \ln 2$$

$$\Sigma [M^2 f(M)]_i = 406,533.96$$

By the trapezoidal rule

$$M_w = 140,661 \simeq \frac{0.693}{2} \times 406,533.96$$

Scholte obtained   141,000 from the MWD
141,000 from experiments at different speeds
142,000 from measurements at one speed

Sample: Polyethylene L-30-76. Solvent: Biphenyl. Temperature: 123.2°C (theta temperature).

Data taken from Scholte's papers (18, 19).

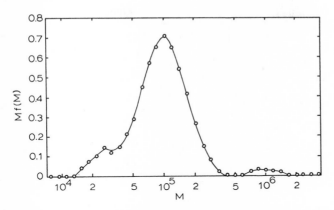

FIG. 8. Scholte's method for MWDs. This is a plot of $Mf(M)$ vs M for poly-ethylene dissolved in biphenyl at the theta temperature (123.2 °C). Redrawn from the data obtained by Scholte (*18, 19*).

Thus the area under the curve for the plot of $f(M)$ vs M must be 1. Clearly we cannot plot f_i vs M_i, even if the intervals between successive molecular weights, ΔM, is constant, since the area under this curve would be ΔM. We could plot $f_i/\Delta M$ (for one series) vs ΔM or $f_i/4\Delta M$ vs ΔM (for data from all four series). But, now note that the interval between successive molecular weights is $2^{1/4}$, i.e., $M_j = M_i 2^{1/4}$. Therefore, Scholte (*18–20*) suggested that the following procedure be used.

$$\sum_i \frac{f_i}{4} \text{(all four series)} = 1 = \int_0^\infty f(M)\, dM$$

$$= \int_0^\infty M f(M) \frac{dM}{M} \simeq \Delta \ln M \sum_i [M f(M)]_i \qquad (94)$$

Since $\Delta \ln M_i$ is constant and equal to $(1/4) \ln 2$ or $0.693/4$, then

$$\sum_i [M f(M)]_i = \frac{4}{0.693} \qquad (95)$$

Since $\sum_i f_i$ (all four series) $= 4$, one notes that for all four series

$$\sum_i \frac{f_i}{0.693} = \frac{4}{0.693} \qquad (96)$$

Thus

$$[M f(M)]_i = f_i/0.693 \qquad (96a)$$

So, if one plots $Mf(M)$ vs $\ln M$, where $\Delta \ln M = \ln M_j - \ln M_i = 0.693/4$, then the area under the curve is 1. This is how Scholte obtained the curve shown in Fig. 8.

Nonsector-Shaped Centerpieces (24, 43)

The starting equation here is Eq. (82), which can be written as

$$V(\Lambda, \xi)_j = \sum_i \frac{\Lambda_j M_i f_i \exp(-\Lambda_j M_i \xi)}{\int_0^1 [\exp(-\Lambda_j M_i \xi)]\, dx} + \delta_j$$
$$= \sum_i f_i H_{ij} + \delta_j \tag{97}$$

where

$$H_{ij} = \frac{\Lambda_j M_i \exp(-\Lambda_j M_i \xi)}{\int_0^1 [\exp(-\Lambda_j M_i \xi)]\, dx} \tag{98}$$

and δ_j is the experimental error. Clearly one can set up an array of $V(\Lambda, \xi)$ in the same manner as one does with the $U(\Lambda, \xi)$ (see Eqs. 91), choose a range of molecular weights, and solve for the f_i's by linear programming. Thus the analysis is done in the same manner as before. Note that for each choice of M_i, the integral in Eq. (98) must be evaluated numerically; the easiest way to do this is to use a computer.

Scholte has also modified his method so that instead of minimizing the sum of the absolute value of the error, one minimizes the square of the error. Results using both procedures give excellent agreement (20).

The Method of Gehatia and Wiff (23, 45–49)

Gehatia and Wiff have developed a very elegant method of determining MWDs from a single sedimentation equilibrium experiment; they claim to be able to analyze multimodal distributions. Figure 9 shows the results obtained with a simulated MWD; the solid line represents the true value and the circles represent the computed distribution. Their method was originally developed for ideal, dilute solution conditions and for sector-shaped centerpieces. Their starting equation is Eq. (61), which they write as (23)

$$U(\xi) = \int_0^\infty K(\xi, M) f(M)\, dM = \int_0^{M_{\max}} K(\xi, M) f(M)\, dM \tag{99}$$

Here

$$K(\xi, M) = \frac{\Lambda M \exp(-\Lambda M \xi)}{1 - \exp(-\Lambda M)} \tag{100}$$

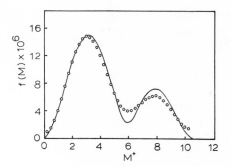

FIG. 9. The Gehatia and Wiff method for MWDs. This is a simulated example (23). The $M^+ = M/13{,}953$. The solid curve represents the true distribution. The open circles represent calculated values of $f(M)$ vs M^+.

M_{max} is the largest molecular weight in the MWD; beyond M_{max}, $f(M) = 0$. They point out that equations of the type shown by Eq. (99) are improperly posed problems in the Hadamard sense. What this means is that small errors in $U(\xi)$ can cause severe oscillations when one attempts to obtain $f(M)$ from this integral equation. A way to avoid this problem is to use a regularizing function which dampens out the oscillations. Although their model was originally set down for cells with sector-shaped centerpieces, it could also be applied to cells with nonsector-shaped centerpieces. In this case Eq. (83) would be used, and it would be written as

$$V(\xi) = \int_0^\infty H(\xi, M)f(M)\, dM = \int_0^{M_{max}} H(\xi, M)f(M)\, dM \qquad (101)$$

where

$$H(\xi, M) = \frac{\Lambda M \exp(-\Lambda M \xi)}{\int_0^1 [\exp(-\Lambda M \xi)]\, dM} \qquad (102)$$

The Gehatia-Wiff method need not be restricted to ideal, dilute solution conditions. Our methods for correcting for nonideal behavior, or a method proposed by Gehatia and Wiff (45, 49) for nonideal solutions, could be used. Their method does require the use of a computer. For more details the reader should consult the papers by Gehatia and Wiff (23, 45–49).

Molecular-Weight Distributions from Sedimentation Velocity Experiments

The sedimentation velocity experiment can also be used to obtain MWDs. Much of the pioneering work in the determination of MWDs from sedimentation velocity experiments was done by Prof. J. W. Williams of the University of Wisconsin-Madison and his associates (4, 5, 24, 52). The advantage of the technique is its rapidity; it takes about 2 hr to do one sedimentation velocity experiment. The disadvantages of the method are (1) its tediousness, (2) the fact that the theory is more empirical and not on as rigorous a foundation as the sedimentation equilibrium method, and (3) that two (sometimes three) extrapolations are involved. Nevertheless, there is an extensive literature on MWDs from sedimentation velocity experiments or on differential distributions of sedimentation coefficients, $g(s)$, which can be transformed to MWDs if a relation between s and M is known (4, 5, 24, 52–57). With the advent of automatic plate readers (58), the tediousness associated with this method can be alleviated, and the availability of pulsed lasers and multiplexers may allow three to five (depending on the rotor type) experiments to be performed simultaneously.

The differential distribution of sedimentation coefficients, $g(s)$, is defined by (4, 5, 43, 52, 53)

$$g(s) = \frac{1}{c}\frac{dc}{ds} = \frac{dG(s)}{ds} \tag{103}$$

Here $G(s)$ is the integral distribution of sedimentation coefficients; it is defined by (4, 5, 43, 52, 53)

$$G(s) = \int_0^s g(s)\,ds = \sum_i r^2 \tilde{n}_{0i}/r_m^2 \tilde{n}_0 \tag{104}$$

The quantity $\tilde{n}_0 = n - n_0$ is the refractive index difference between solution (n) and solvent (n_0); $\tilde{n}_{0i}$ is the refractive index difference for component i. In these equations c is the concentration of the macromolecular solutes, s is the sedimentation coefficient, r is the radial position in the moving boundary, and r_m is the radial position of the air–solution meniscus. If the refractive index increments, ψ_i, of the macromolecular components are all the same, then $n - n_0 = \psi c$, and $G(s)$ becomes

$$G(s) = \sum_i r^2 c_{0i}/r_m^2 c_0 \tag{105}$$

Equation (103) for $g(s)$ is often written in a more useful form, namely, (4, 5, 43, 52, 53)

$$g(s) = \frac{r^3 \omega^2 t}{c_0 r_m^2}\left(\frac{\partial c}{\partial r}\right)_t \tag{106}$$

Here we have used the relations

$$c = c_0(r_m/r)^2 \tag{107}$$

$$\frac{dc}{ds} = \frac{dc}{dr}\frac{dr}{ds} \tag{108}$$

and

$$\frac{dr}{ds} = \frac{d}{ds}(r_m \exp[s\omega^2 t])$$
$$= t\omega^2 r_m \exp(s\omega^2 t) = r\omega^2 t \tag{109}$$

Since both sedimentation and diffusion occur in a moving boundary, the experimentally measured $g(s)$ is actually an apparent value, $g^*(s)$, which is also defined by Eq. (106). The spreading of the moving boundary due to sedimentation (s) is proportional to t, while the spreading of the boundary due to diffusion (D) is proportional to $t^{1/2}$ (5, 59). If the diffusion effects are not too great, there should be an intermediate region of time in which values of $g^*(s)$ and $1/t$ are linearly correlated. Thus extrapolation of $g^*(s)$ to $1/t = 0$ should eliminate diffusion effects; values of $g^*(s)$ at $1/t = 0$ are designated by $g°(s)$. With nonaqueous solutions, pressure effects are important, and a correction must also be made for them. For details on these corrections, consult the reviews by Williams et al. (4), Baldwin and Van Holde (53), or the monograph by Fujita (5).

Once $g°(s)$ has been obtained, it is necessary to remove concentration dependence so that the true $g(s)$ can be obtained. There are several ways to solve this problem(s). In one method, introduced by Baldwin et al. (3, 60, 61), values of $g°(s)/g°_{max}(s)$ at infinite time are obtained for each boundary gradient curve at a certain initial concentration c_0. Here $g°_{max}(s)$ is the maximum value of $g(s)$. Then one extrapolates $1/s$ to c_0 at each fixed value of $g°(s)/g°_{max}(s)$ for all of the sedimentation velocity experiments at various initial concentrations. This procedure gives s_0, the sedimentation coefficient at zero concentration, for each ratio of $g°(s)/g°_{max}(s)$. One obtains $1/g°_{max}(s)$ from (5, 60, 61)

$$1/g°_{max}(s) = \int_0^\infty [g°(s)/g°_{max}(s)]\,ds \tag{110}$$

Once $g°_{max}(s)$ is known, then it is a simple matter to obtain $g°(s_0)$. A plot of $g°(s_0)$ vs s_0 gives the plot of the differential distribution of sedimentation coefficients. The integral distribution of sedimentation coefficients, $G(s_0)$, is obtained from Eq. (104). Figure 10 shows a plot of $g°(s_0)$ vs s_0 and $G°(s_0)$ vs s_0 for a gelatin sample (60).

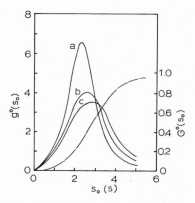

FIG. 10. Differential, $g°(s)$ and integral, $G°(s_0)$, distributions of sedimentation coefficients for a gelatin sample (60). The Gaussian Curves a, b, and c represent values of $g°(s)$ at $c = 0.746$ g/dl (a), at $c = 0.306$ g/dl (b), and at $c = 0$ (c). The integral distribution (— ·), $G°(s_0)$, is obtained by numerical integration of Curve c.

In order to obtain a differential distribution of molecular weights, $f(M)$, one has to have an empirical relation of the form (4, 5, 43)

$$s_0 = KM^\alpha \tag{111}$$

or

$$s_0 = KM_w{}^\alpha \tag{111a}$$

The constants K and α depend on the temperature and the polymer–solvent combination. This relation can be established by measuring the sedimentation coefficients and molecular weights (M) or weight-average molecular weights (M_w) of some polymer fractions or some samples of the same type polymer. Values of K and α for many polymers are tabulated in the *Polymer Handbook* (14). If one lets $g°(s_0)\, ds_0$ be the weight fraction of polymer having a sedimentation coefficient (at infinite dilution) between s_0 and $s_0 + ds_0$, then because of Eqs. (111) and (111a) one notes that

$$g°(s_0)\, ds_0 = f(M)\, dM \tag{112}$$

and

$$f(M) = g°(s_0)\frac{ds_0}{dM} = \frac{\alpha s_0 g°(s_0)}{M} \tag{113}$$

Figure 11 shows a plot of $f(M)$ vs M for a trimodal polystyrene sample obtained at the theta temperature, 34.2 °C, in cyclohexane by sedimentation equilibrium (top) and by sedimentation velocity (bottom) experiments. Note the excellent agreement in the two techniques; the data are taken from the paper by Scholte (*20*). Table 2 shows the values of M_n, M_w, and M_z obtained by the two ultracentrifugal techniques and by gel permeation chromatography (*20*). The agreement between the methods is quite good, demonstrating the versatility of the ultracentrifuge in this area.

Concluding Remarks

By far the most commonly used method for obtaining a MWD is the gel permeation method, which was developed by Moore (*62*) for the rapid determination of MWDs of thermoplastics. Many more recent details about gel permeation chromatography will be found in the book edited by Ezrin (*63*). This book contains the proceedings of a conference on Polymer

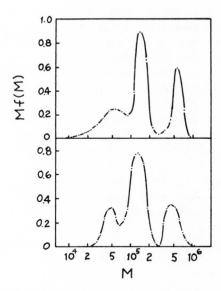

FIG. 11. Plots of $Mf(M)$ vs M for a trimodal blend of polystyrene. The upper curve was obtained from sedimentation velocity experiments using $g°(s_0)$; the lower curve was obtained from sedimentation equilibrium experiments using Scholte's linear programming method (*20*). Values of M_n, M_w, and M_z for this sample obtained by various methods are listed in Table 2.

TABLE 2
Average Molecular Weights of a Blend of Three Polystyrene Samples

How obtained	$M_n \times 10^3$	$M_w \times 10^3$	$M_z \times 10^3$
1. From average molecular weights of the original samples	102	204	385
2. From sedimentation equilibrium experiments by Scholte's method (Eq. 59) using linear programming	106	202	362
3. From sedimentation velocity experiments [$g(s)$]	101	205	365
4. From gel permeation chromatography	96	204	439

Molecular Weight Methods; about one-third of the papers in the book deal with gel permeation chromatography. We have seen in Fig. 7 that there is good agreement with the MWD of the dextran sample determined from sedimentation equilibrium experiments (43, 44) and by a combination of analytical gel chromatography and light scattering. The details of the manufacturer's method for obtaining the MWD of the dextran sample are given in the papers by Granath (64, 65). We have also seen in Table 2 that there is good agreement between ultracentrifugal methods for obtaining MWDs and gel permeation chromatography. The agreement between the various methods is gratifying.

The ultracentrifuge has been used for the characterization of latex particles (66, 67). Electron microscopy established that the particles were spherical. The latex particles are large compared to the wavelength of the blue ($\lambda = 456$ nm) or the green ($\lambda = 546$ nm) lines of mercury, so that the particles undergo Mie scattering. Thus a photoelectric scanner could be used to obtain sedimentation coefficients, s, and $g°(s_0)$, the differe ntial distribution of sedimentation coefficients. From the Stokes-Einstein relation s could be related to x^2, the square of the radius of the particles; hence, a distribution of radii could also be obtained. Here is a beautiful example of a practical application of the ultracentrifuge, since it could be used for quality control.

Wales and Rehfeld (55) have related the intrinsic viscosity, [η], and $g°(s_0)$ for linear polymers. Their procedure has been used by Merle and Sarko (56) and by Bluhm and Sarko (57) to obtain a polydispersity index, i.e., a ratio of M_w/M_n, in addition to $g(s_0)$ for some synthetic, linear, stereo-regular polysaccharides.

Automatic plate readers have been developed for reading Rayleigh fringes (58). Modulatable lasers are available now; in fact we are using them ourselves. Multihole rotors (4 to 6 holes) are also available so that 3 to 5 sedimentation velocity or 3 to 15 sedimentation equilibrium experiments (3 to 5 for a sector-shaped centerpiece or 9 to 15 for a multichannel equilibrium centerpiece) can be performed in the same time it takes to perform one experiment. An interferometric optical system which gives refractive index gradients as a family of fringes has been developed by Bryngdahl and Ljunggren (68); it is used on the Christ Omega II ultracentrifuge. The patterns produced by this optical system (68), the Rayleigh optical system, and perhaps the schlieren optical system could be analyzed with automatic plate readers. The output from the plate reader could be fed into a computer. Thus the tediousness associated can be removed. Ultracentrifugal analysis for heterogeneity, homogeneity, or MWDs should have an interesting future.

SELF-ASSOCIATIONS

Introduction

Chemical equilibria of the type

$$nP_1 \rightleftarrows P_n, \quad n = 2, 3, \ldots \tag{114}$$

$$nP_1 \rightleftarrows qP_2 + mP_3 + \cdots \tag{115}$$

and related equilibria involving a solute P are known as self-associations. These reactions are widely encountered. Among materials that undergo self-association are many proteins (30–32, 35, 40), soaps and detergents (69–72), purine (33, 73), and nucleosides and nucleotides (33, 74, 75) as well as some polymers (76). The existence of self-associations was recognized early in the development of the ultracentrifuge (3). The first theoretical treatment of self-associations by sedimentation equilibrium was by Tiselius (77); no methods for analyzing self-associations were presented in his paper. The next development was by Debye who showed how the micellar aggregation of soaps and detergents could be studied by light scattering (69); this treatment can be applied to sedimentation equilibrium experiments. Steiner developed a very elegant method for analyzing self-associations by light scattering (78) or by osmotic pressure (79). Some of his procedures were applied to the Archibald method (an ultracentrifugal technique related to sedimentation equilibrium experiments) by Rao and Kegeles (80) and were used by them to study the self-association of α-chymotrypsin in phosphate buffer.

Up to this point the analysis of self-associations was restricted to ideal, dilute solutions. Adams and Fujita (81) first showed how to analyze a nonideal self-association; their treatment was restricted to a monomer–dimer association. Adams and Williams (82) extended the analysis to monomer–n-mer associations beyond dimer ($n \geq 3$); they introduced the apparent weight fraction of monomer, f_a, into the analysis. Subsequently Adams (29) discovered an interrelation between the number- (M_{nc}) and weight- (M_{wc}) average molecular weights (or their apparent values in nonideal solutions). He showed how other self-associations besides monomer–n-mer associations could be analyzed. Since then new improvements in the method of analyzing self-associations have been developed. Chun et al. (34) introduced a graphical procedure for studying nonideal monomer–n-mer and indefinite self-associations; their method makes the analysis much neater and simpler. Another method for analyzing self-associations was developed by Derechin (83, 84) who used the multinomial theorem. More recently Šolc and Elias (85) have studied the theory for heterogeneous self-associations; this is an area still in its infancy.

In the study of self-associations one is often dealing with multicomponent systems. With an associating protein one may need a buffer to control the pH, plus some supporting electrolytes, such as NaCl or KCl, to swamp out charge effects. Thus one has three or more components—the associating solute, the supporting electrolyte, water, and sometimes buffers. Vrij and Overbeek (86) and also Casassa and Eisenberg (87) showed that if a solution containing an ionizable, macromolecular solute was dialyzed against a solution containing supporting electrolyte and/or buffers, then the sedimentation equilibrium, light-scattering, or osmotic pressure equations reduce to a form that is formally identical to the equations for a two-component system. Heretofore most studies on ionizable, self-associating solutes have been restricted to larger molecules such as proteins, but the recent development of hollow fiber dialyzers with a low molecular weight cut-off (200 Daltons) now allows the study of small, ionizable solutes. In this section the equations for a two-component system will be used. The equilibrium constant(s), K_i, the second virial coefficient, BM_1, the partial specific volume, $\bar{v}$, and the refractive index increment, ψ, refer to constituents defined by the Vrij-Overbeek (86) or Casassa-Eisenberg (87) conventions.

Conditions for Simultaneous Chemical and Sedimentation Equilibrium

At constant temperature the condition for chemical equilibrium for reactions described by Eqs. (114) and (115) is (30, 81)

$$n\mu_1 = \mu_n, \qquad n = 2, 3, \ldots \tag{116}$$

Here μ_i ($i = 1, 2, \ldots$) is the molar chemical potential of associating species i. The condition for sedimentation equilibrium requires that the total potential, $\bar{\mu}_i$, for each constituent be constant everywhere in the cell. $\bar{\mu}_i$ is defined by Eq. (1). For self-associations one notes that

$$\begin{aligned} M_2 &= 2M_1 \\ M_3 &= 3M_1 \end{aligned} \tag{117}$$

or

$$M_j = jM_1$$

Using Eqs. (1), (116), and (117), it can be shown for self-associating solutes that

$$n\bar{\mu}_1 = \bar{\mu}_n = \text{constant} \tag{118}$$

at sedimentation equilibrium ($n = 2, 3, \ldots$). This means that the total molar potential of constituent i has the same relation that the chemical potential has when self-association is present.

Assumptions Required for the Analysis of Self-Associations

In order to analyze self-associations it is necessary to make the following assumptions (29–35, 81): (a) The partial specific volumes, $\bar{v}$, of all the associating solutes are the same, or the density increments, $(\partial\rho/\partial c)_\mu$, are the same for all associating solutes. (b) The refractive index increments, ψ, of the associating species are equal. (c) The natural logarithm of the activity coefficient, y_i, on the c-scale (grams/liter) can be represented by

$$\ln y_i = iBM_1c, \qquad i = 2, 3, \ldots \tag{119}$$

Here B is a constant that is characteristic of the solute–solvent mixture; BM_1 is known as the second virial coefficient. It has been shown in light-scattering experiments on mercaptalbumin and its mercury dimer that Eq. (119) is valid (88). Furthermore, in his study of the self-association of organic dyes, Braswell (89) pointed out that in its limiting form the Debye-Hückel theory indicates that the mean ionic activity coefficients have a similar relation, i.e.,

$$\gamma_\pm(\text{dimer}) = \gamma_\pm{}^2(\text{monomer}) \tag{120}$$

Another reason for using Eq. (119) is that it makes the analysis easier, since K_i and BM_1 can be evaluated from experimentally available data. If the ψ and $\bar{v}$ or $(\partial\rho/\partial c)_\mu$ differ for each constituent, then one does not obtain M_{wc} and the analysis becomes more formidable.

As a consequence of Eq. (119), the concentration of species n can be expressed as (29–35, 81)

$$c_n = K_n c_1{}^n \tag{121}$$

and the total concentration, c, of the associating solute is given by

$$c = c_1 + K_2 c_1{}^2 + \cdots + K_j c_1{}^j + \cdots \tag{122}$$

The second virial coefficient, BM_1, does appear in the expressions for the apparent average molecular weights and for the apparent weight fraction of monomer.

It has been shown that one cannot use $M_{w\,\text{cell}}$ (see Eq. 45) in the analysis of self-associations; instead the quantity M_{wr} (see Eq. 30) must be used. Adams and Fujita (81) showed that M_{wr} or its apparent value, $M_{wr\,\text{app}}$, are functions of the total solute concentration, c, for self-associations. Thus the symbols M_{wc} and M_{wa} will be used in place of M_{wr} and $M_{wr\,\text{app}}$; the subscript c will indicate that we are dealing with a self-association. For self-associating solutes the basic sedimentation equilibrium equation becomes (29–35, 81)

$$\frac{d\ln c}{d(r^2)} = AM_{wa} = \frac{AM_{wc}}{(1 + BM_{wc}c)} \tag{123}$$

Here

$$A = (1 - \bar{v}\rho)\omega^2/2RT \tag{124}$$

or

$$A = 1000(\partial\rho/\partial c)_\mu \omega^2/2RT \tag{125}$$

and

$$\frac{M_1}{M_{wa}} = \frac{M_1}{M_{wc}} + BM_1 c \tag{126}$$

For an ideal, dilute solution, $BM_1 = 0$ and $M_{wa} = M_{wc}$.

In order to analyze self-associations it is necessary to do experiments with a series of solutions of different initial concentrations, c_0. Sedimentation equilibrium experiments are performed on each solution, and for each solution one can obtain values of M_{wa} vs c ($c = c_r$). These values of M_{wa} vs c for each experiment are pieced together to make a plot of M_{wa} vs c as shown in Fig. 12. Here the different symbols indicate results obtained

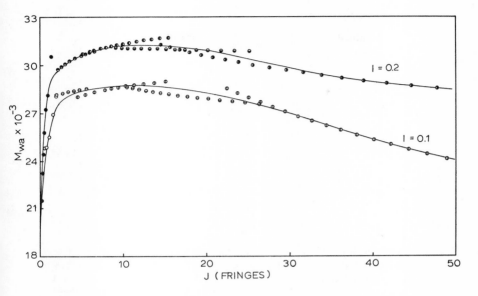

FIG. 12. Self-association of β-lactoglobulin C in glycine buffers at 10 °C. Both buffers had 0.2 M glycine and 0.1 M HCl (pH 2.46 at 23.5 °C). The second buffer had 0.1 M KCl in addition so that its ionic strength was 0.2. In both cases nonideal behavior is observed, and a monomer–dimer association appears to be present (104). At 25 °C, $J = 3.394c$ for c in g/l ($\lambda = 632.8$ nm; 12 mm centerpiece).

with solutions of different initial concentrations; remember that $M_{wa} = f(c)$ for self-associations. A smooth curve is drawn through the M_{wa} vs c plot, and the plot is extrapolated to M_1, the monomer molecular weight. In principle one should be able to extrapolate values of M_{wa} or $M_{wa}(1 - \bar{v}\rho)$ to zero concentration so that the correct value of M_1 or $M_1(1 - \bar{v}\rho_0)$ required by the Vrij-Overbeek (86) or Casassa-Eisenberg (87) conventions is obtained. In actual practice one is forced to choose a value of M_1 from amino acid analysis or some other method, since with strong associations the plots of M_{wa} vs c or $1/M_{wa}$ vs c become quite steep in the vicinity of zero concentration.

Relation Between M_{wa}, M_{na}, and $\ln f_a$. Applicability to Light Scattering and Osmometry

From the smooth plot of M_{wa} vs c a plot of M_1/M_{wa} vs c can be constructed. A minimum in this plot (or a maximum in the plot of M_{wa} vs c)

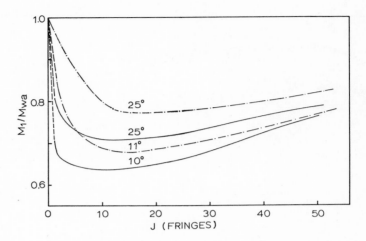

FIG. 13. Self-association of β-lactoglobulins A (— ·) and C(——) in 0.2 M glycine buffer (pH 2.46 at 23.5 °C; ionic strength 0.1). Note the temperature effect on the self-association. Also note that β-lactoglobulin C undergoes a stronger monomer–dimer self-association than does the A variant. Both proteins exhibit nonideal behavior under these solution conditions. These genetic variants differ by three amino acids. The concentrations are given in red fringes (12 mm centerpieces; $\lambda = 632.8$ nm); at 25 °C, $J = 3.394c$ for c in g/l (35, 104).

is indicative that a nonideal self-association is present. Figure 13 shows such a plot. From plots of M_1/M_{wa} vs c one can obtain M_{na} (29, 30), the apparent number-average molecular weight, and $\ln f_a$ (30, 82), the natural logarithm of the apparent weight fraction of monomer. The quantity M_1/M_{na} is obtained from

$$\frac{M_1}{M_{na}} = \frac{1}{c} \int_0^c \frac{M_1}{M_{wa}} \, dc \tag{127}$$

where

$$\frac{M_1}{M_{na}} = \frac{M_1}{M_{nc}} + \frac{BM_1 c}{2} \tag{128}$$

The quantity $\ln f_a$ is obtained from

$$\ln f_a = \int_0^c \left(\frac{M_1}{M_{wa}} - 1 \right) \frac{dc}{c} \tag{129}$$

where

$$\ln f_a = \ln f_1 + BM_1 c \tag{130}$$

and

$$f_1 = c_1/c \tag{131}$$

is the weight fraction of monomer. Note that the

$$\lim_{c \to 0} \left(\frac{M_1}{M_{wa}} - 1 \right) \bigg/ c$$

exists, so that the lower limit of the integral is zero. If dimer is present, the limit is $-K_2 + BM_1$, and if dimer is absent the limit is BM_1. Since M_{wa} can be obtained from light-scattering experiments, it follows that one can also obtain M_{na} and $\ln f_a$ from light-scattering experiments provided one has done enough experiments to obtain a plot of M_{wa} vs c. Each value of M_{wa} corresponds to one light scattering or one osmotic pressure experiment. So it is evident that one can obtain far more information from a series of sedimentation equilibrium experiments on a few solutions; for instance, five or more different initial concentrations were required to obtain the data shown in Fig. 12 and 13. Since M_{na} can be obtained from osmotic pressure experiments, one can construct a plot of M_{na} vs c if several experiments are performed. With this information M_{wa} and $\ln f_a$ can be obtained since (90)

$$\frac{M_1}{M_{wa}} = \frac{d}{dc} \left(\frac{cM_1}{M_{na}} \right) = \frac{M_1}{M_{na}} + c \frac{d}{dc} \left(\frac{M_1}{M_{na}} \right) \tag{132}$$

and because of Eq. (132)

$$\ln f_a = \int_0^c \left(\frac{M_1}{M_{na}} - 1 \right) \frac{dc}{c} + \left(\frac{M_1}{M_{na}} - 1 \right) \tag{133}$$

It should be apparent that if a physical method gives an average molecular weight or its apparent value at various concentrations for a self-associating solute, then it is potentially possible to analyze the self-association by the methods described here.

It is possible to eliminate the second virial coefficient, BM_1, by various combinations of M_1/M_{na}, M_1/M_{wa}, and $\ln f_a$; the resulting equations are particularly useful for the analysis of monomer–n-mer and indefinite self-associations. Equations (126) and (128) can be combined to give (34, 35)

$$\xi = \frac{2M_1}{M_{na}} - \frac{M_1}{M_{wa}} = \frac{2M_1}{M_{nc}} - \frac{M_1}{M_{wc}} \tag{134}$$

Similarly, Eqs. (126) and (130) can be combined to give

$$\eta = \frac{M_1}{M_{wa}} - \ln f_a = \frac{M_1}{M_{wc}} - \ln f_1 \tag{135}$$

Equations (128) and (130) can be combined to give

$$v = \frac{2M_1}{M_{na}} - \ln f_a = \frac{2M_1}{M_{nc}} - \ln f_1 \tag{136}$$

These relations were developed by Chun et al. (34). The first two relations are most useful in sedimentation equilibrium experiments (35). The third relation is most useful in osmotic pressure experiments. Note that BM_1 has been eliminated in Eqs. (134)–(136), and that the quantities ξ, η, and v have the same values they would have under ideal conditions.

Analysis of a Monomer-n-Mer Association

The association being considered is described by Eq. (114). For this association one notes that the following relations apply (34, 35, 91):

$$c = c_1 + K_n c_1^n \tag{137}$$

$$1 = f_1 + f_n \quad \text{or} \quad f_n = 1 - f_1 \tag{138}$$

$$f_n = K_n c_1^n / c = K_n c^{n-1} f_1^n \tag{139}$$

The quantities M_1/M_{nc} and M_1/M_{wc} are defined as

$$\frac{M_1}{M_{nc}} = \frac{1}{c}\left(c_1 + \frac{K_n c_1^n}{n}\right)$$

$$= f_1 + \frac{f_n}{n} = \frac{1 + f_1(n - 1)}{n} \tag{140}$$

and

$$\frac{M_1}{M_{wc}} = \frac{c}{c_1 + nK_n c_1^n} = \frac{1}{n + f_1(1 - n)} \tag{141}$$

so that

$$\xi = \frac{2M_1}{M_{nc}} - \frac{M_1}{M_{wc}} = \frac{2 - 2f_1(1 - n)}{n} - \frac{1}{n + f_1(1 - n)} \tag{142}$$

Equation (142) is quadratic in f_1, hence f_1 is given by

$$f_1 = \frac{n}{4(n - 1)^2}\left\{(n - 1)\left(\xi + 2 - \frac{2}{n}\right) - \left(\left[(n - 1)\left(\xi + 2 - \frac{2}{n}\right)\right]^2\right.\right.$$

$$\left.\left. - \left(\frac{8}{n}\right)(n - 1)^2(\xi n - 1)\right)^{1/2}\right\} \tag{143}$$

Thus once f_1 is known, it is a simple matter to use Eqs. (138) and (139) to obtain K_n, since

$$\frac{1 - f_1}{f_1} = K_n(cf_1)^{n-1} \tag{144}$$

A plot of $(1 - f_1)/f_1$ vs $(cf_1)^{n-1}$ has a slope equal to K_n. Once f_1 is known, M_1/M_{wc} is known (see Eq. 141), so that Eq. (126) can be rewritten as

$$\frac{M_1}{M_{wa}} - \frac{1}{n + f_1(1 - n)} = BM_1c \tag{145}$$

A plot of

$$\frac{M_1}{M_{wa}} - \frac{1}{n + f_1(1 - n)} \text{ vs } c$$

has a slope of BM_1. So far it has been assumed n is known. If n is unknown, then one must assume values of n ($n = 2$, 3, etc.) and solve for f_1, K_n, and BM_1 for each choice. The correct choice will give straight line plots which pass through or close to origin. Figure 14 shows an example of a test for

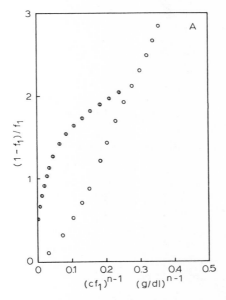

FIG. 14a. Test for a monomer–n-mer self-association using $n = 2$ and $n = 3$. β-Lactoglobulins A and C in 0.2 M glycine buffer (0.1 ionic strength, pH 2.46 at 23.5 °C). Results at 11 °C with β-lactoglobulin A. $K_2 = 9.58$ dl/g (35). Values of f_1 used here were calculated from ξ (see Eq. 142). The plots for $n = 2$ come closest to describing a straight line that passes close to the origin; this indicates the presence of a monomer–dimer association. Attempts to analyze these associations as a monomer–n-mer association with $n > 3$ were unsuccessful.

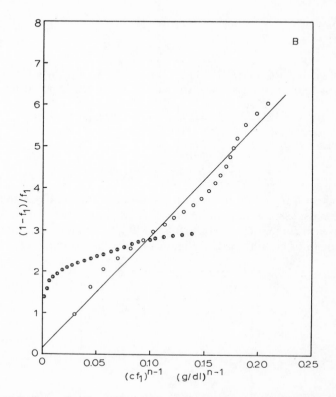

FIG. 14b. Same as Fig. 14a, but these results are at 10°C with β-lactoglobulin C. $K_2 = 27.2$ dl/g (104).

a monomer–n-mer association; note that the plot for $n = 2$ gives straight lines which come closest to satisfying plots based on Eqs. (144) and (145). It is also possible to use η (see Eq. 135) to try to analyze a monomer–n-mer association; for this case

$$\eta = \frac{M_1}{M_{wa}} - \ln f_a = \frac{1}{n + f_1(1 - n)} - \ln f_1 \qquad (146)$$

This equation has one unknown, f_1, which is solved for by successive approximations; remember that $0 < f_1 \le 1$. A plot based on Eq. (144) using values of f_1 obtained from Eq. (146) is also shown in Fig. 15. For the sedimentation equilibrium experiment it has been shown that the quantity ξ (see Eq. 142) is the least affected by experimental error of the

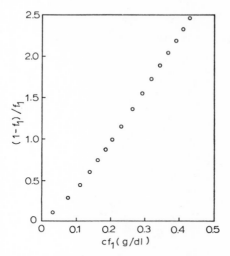

FIG. 15. Self-association of β-lactoglobulin A at 11 °C in 0.2 M glycine buffer (ionic strength 0.1, pH 2.46 at 23.5 °C). Test for the presence of a monomer–dimer association using values of f_1 calculated from η (see Eq. 146). From the slope of the best straight line through this plot, a value of $K_2 = 6.56$ dl/g was obtained. Note the difference in this value and the one obtained in Fig. 14a. The plot based on Eq. (142) has been shown to be more reliable (*35*).

three quantities, ξ, η, and v. On the other hand both ξ and η, being functions of f_1 only, can be used to set up standard plots of ξ vs η for various choices of n. This means experimental values of ξ and η could be calculated, and these calculated values could be compared to a standard table or plot of ξ vs η to see if a monomer–n-mer association is present. Figure 16 shows such a standard plot for a few values of n.

Analysis of Indefinite Self-Associations

A sequential indefinite self-association is described by Eq. (115). Here the association appears to continue without limit. For ideal, dilute solutions or for nonideal solutions for which Eq. (122) applies, one can represent an indefinite self-association as being made up of simultaneous associations of the type (*30, 32, 33, 76*)

$$P_1 + P_1 \rightleftarrows P_2, \qquad K_{12} = [P_2]/[P_1]^2$$
$$P_1 + P_2 \rightleftarrows P_3, \qquad K_{23} = [P_3]/[P_1][P_2] \qquad (147)$$
$$P_1 + P_3 \rightleftarrows P_4, \qquad K_{34} = [P_4]/[P_1][P_3]$$

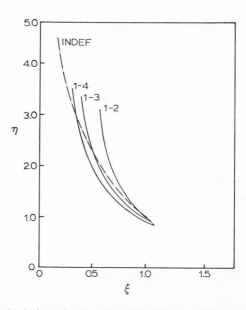

FIG. 16. Standard plots of η (see Eqs. 142 and 161) vs ξ (see Eqs. 146 and 162) for a monomer–dimer (1–2), a monomer-trimer (1–3), a monomer-tetramer (1–4), and a sequential, indefinite (INDEF) self-association. For these associations η and ξ are each functions of f_1 ($0 \le f_1 \le 1$); thus, for each model, one can assume values of f_1 and construct standard curves which can be used to test for the type of self-association that might be present (*34*).

and so on. Here $[P_i]$ represents the molar concentration of species i ($i = 1, 2, \ldots$), and K_{ij} represents the molar equilibrium constant. The Eqs. (147) can be rearranged to give

$$[P_2] = K_{12}[P_1]^2$$
$$[P_3] = K_{12}K_{23}[P_1]^3 \tag{148}$$
$$[P_4] = K_{12}K_{23}K_{34}[P_1]^4$$

and so on. In order to analyze an indefinite self-association, one must make some assumptions regarding the molar equilibrium constants, the K_{ij}; otherwise, the analysis becomes formidable. For a sequential indefinite self-association (an association in which all species appear to be present), one can assume that all molar equilibrium constants are equal, i.e., assume

$$K_{12} = K_{23} = K_{24} = \cdots = K \tag{149}$$

Now convert concentrations to the gram/milliliter scale, C_i; thus we obtain

$$C_2 = 2\left(\frac{1000K}{M_1}\right)C_1{}^2 = 2kC_1{}^2$$

$$C_3 = 3\left(\frac{1000K}{M_1}\right)^2 C_1{}^3 = 3k^2C_1{}^3 \tag{150}$$

and so on. The total concentration of the associating solute becomes

$$\begin{aligned}
C &= C_1 + 2kC_1{}^2 + 3k^2C_1{}^3 + 4k^3C_1{}^4 + \cdots \\
&= C_1(1 + 2kC_1 + 3k^2C_1{}^2 + 4k^3C_1{}^3 + \cdots) \\
&= C_1/(1 - kC_1)^2, \quad \text{if } kC_1 < 1
\end{aligned} \tag{151}$$

Under these circumstances Eq. (119) becomes

$$\ln y_i = i\hat{B}M_1C = iBM_1c \tag{119a}$$

Since $C = c/1000$, then $\hat{B}M_1 = 1000BM_1$. Note that M_1/M_{wa} can be written as

$$\frac{M_1}{M_{wa}} = \frac{M_1}{M_{wc}} + \hat{B}M_1C \tag{152}$$

Now M_1/M_{wa} is the same whether the gram/liter or the gram/milliliter concentration scale is used; this follows from Eq. (123) since $d \ln c = d \ln C$. Thus it follows that

$$\hat{B}M_1C = BM_1c \tag{153}$$

The number-average molecular weight, M_{nc}, is defined as

$$M_n = \sum_i n_i M_i / \sum_i n_i = w/\sum_i w_i/M_i$$

This definition can be recast as

$$M_{nc} = C/\sum_i (C_i/M_i)$$

It follows that

$$\frac{C}{M_{nc}} = \sum_i \frac{C_i}{M_i}$$

For a self-associating system $M_2 = 2M_1$, $M_3 = 3M_1$, etc., so that

$$\begin{aligned}
\frac{CM_1}{M_{nc}} &= M_1 \sum_i \frac{C_i}{M_i} = \sum_i \frac{C_i}{i} \\
&= C_1 + kC_1{}^2 + k^2C_1{}^3 + k^3C_1{}^4 + \cdots \\
&= C_1/(1 - kC_1), \quad \text{if } kC_1 < 1
\end{aligned} \tag{154}$$

Division of Eq. (154) by Eq. (151) leads to

$$M_1/M_{nc} = 1 - kC_1 \qquad (155)$$

Similarly it can be shown that

$$\frac{M_1}{M_{wc}} = \frac{1 - kC_1}{1 + kC_1} \qquad (156)$$

The expressions for M_1/M_{nc} and for M_1/M_{wc} are formally the same as those obtained by Flory (92) for linear condensation polymerization. Flory obtained $M_0/M_n = 1 - p$ and $M_0/M_w = (1 - p)/(1 + p)$; here M_0 is the molecular weight of the repeating unit and $p(0 \le p \le 1)$ is the extent of polymerization.

Equation (151) can be rearranged to give

$$C_1/C = f_1 = (1 - kC_1)^2 \qquad (157)$$

from which it follows that

$$\sqrt{f_1} = 1 - kC_1 = 1 - kCf_1 \qquad (158)$$

or

$$(1 - \sqrt{f_1})/f_1 = kC \qquad (158a)$$

Once $\sqrt{f_1}$ or f_1 has been obtained, one can use Eq. (158a) to obtain k, since a plot of $(1 - \sqrt{f_1})/f_1$ vs C would be a straight line passing through the origin and having a slope of k. Equation (158) could also be used for this purpose. Experimental error may cause the plot not to go through the origin, but the plot should come close to it. We can use Eqs. (152), (156), and (158) to show that

$$\frac{M_1}{M_{wa}} = \frac{1 - kC_1}{1 + kC_1} + \hat{B}M_1C$$

$$= \frac{\sqrt{f_1}}{2 - \sqrt{f_1}} + \hat{B}M_1C \qquad (159)$$

Similarly, M_1/M_{na} can be expressed as

$$\frac{M_1}{M_{na}} = 1 - kC_1 + \frac{\hat{B}M_1C}{2}$$

$$= \sqrt{f_1} + \frac{\hat{B}M_1C}{2} \qquad (160)$$

For an indefinite self-association the quantities ξ and η (see Eqs. 134 and

135) are given by

$$\xi = 2\sqrt{f_1} - \frac{\sqrt{f_1}}{2 - \sqrt{f_1}} \tag{161}$$

and

$$\eta = \frac{\sqrt{f_1}}{2 - \sqrt{f_1}} - \ln f_1 \tag{162}$$

Equation (161) is quadratic in $\sqrt{f_1}$, and $\sqrt{f_1}$ is obtained from

$$\sqrt{f_1} = (1/4)\{(\xi + 3) - \sqrt{(\xi + 3)^2 - 16\xi}\} \tag{163}$$

One obtains f_1 or $\sqrt{f_1}$ in Eq. (162) by successive approximations. Once f_1 or $\sqrt{f_1}$ is known, k can be obtained from plots based on Eq. (158) or (158a). The second virial coefficient $\hat{B}M_1$ is obtained from Eq. (161) since

$$\frac{M_1}{M_{wa}} - \frac{\sqrt{f_1}}{2 - \sqrt{f_1}} = \hat{B}M_1 C \tag{164}$$

A plot of

$$\left\{\frac{M_1}{M_{wa}} - \frac{\sqrt{f_1}}{2 - \sqrt{f_1}}\right\}$$

vs C will have a slope of $\hat{B}M_1$. This plot should go through the origin, but experimental error may cause it to miss the origin slightly. Figure 17 shows plots based on Eqs. (158a) and (160).

The indefinite self-association considered above is one for which all molar equilibrium constants are equal, and it also implies that the standard Gibbs free energy for the addition of monomer (or unimer) to an aggregate of any size, including monomer, is the same. Now suppose that only the standard Gibbs free energy for the dimerization is different from the standard Gibbs free energy for formation of higher aggregates. This would mean that in Eqs. (147) and (148), $K_{12} \neq K_{23}$, K_{34}, etc., but that $K_{23} = K_{34} = \cdots = K$. In this case it can be shown that (93)

$$C = C_1\left[1 + \frac{k_{12}C_1(2 - kC_1)}{(1 - kC_1)^2}\right] \tag{165}$$

if $kC_1 < 1$. Here

$$k_{12} = 1000K_{12}/M_1 \tag{166}$$

and

$$k = 1000K/M_1 \tag{167}$$

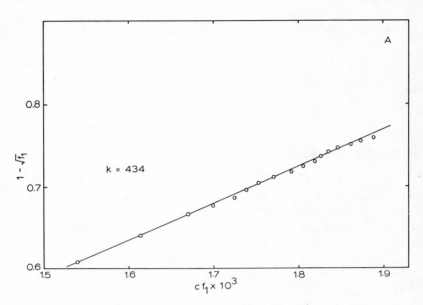

FIG. 17a. Test for a sequential, indefinite self-association. Self-association of β-lactoglobulin A at 16°C in 0.2 M acetate buffer (ionic strength 0.1, pH 4.61 at 22.5°C). Using the data of Adams and Lewis (32), f_1 was calculated from ξ (see Eq. 161) and used for the evaluation of k and $\hat{B}M_1$. Evaluation of k from a plot based on Eq. (158). By another method Adams and Lewis obtained $k = 400$ ml/g and $\hat{B}M_1 = 1.6$ ml/g. Note the amazing effect of the solution conditions (see Fig. 14) on the self-association of the β-lactoglobulin A.

Similarly, it can be shown that

$$\frac{M_1}{M_{wc}} = \frac{\left[1 + \dfrac{k_{12}C_1(2 - kC_1)}{(1 - kC_1)^2}\right]}{\left[1 + \dfrac{k_{12}C_1(4 - 3kC_1 + k^2C_1^2)}{(1 - kC_1)^3}\right]} \tag{168}$$

and that

$$\frac{M_1}{M_{nc}} = \frac{\left[1 + \dfrac{k_{12}C_1}{(1 - kC_1)}\right]}{\left[1 + \dfrac{k_{12}C_1(2 - kC_1)}{(1 - kC_1)^2}\right]} \tag{169}$$

The quantity ξ (see Eqs. 134 and 135) would involve two unknowns, $x = k_{12}C$ and $y = kC_1$, and it would be awkward to solve for these

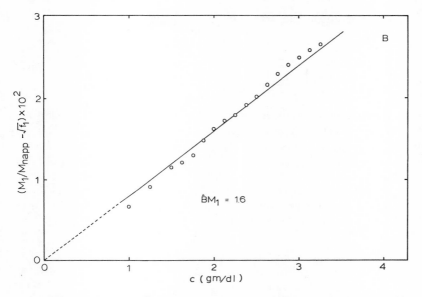

FIG. 17b. Evaluation of $\hat{B}M_1$ from a plot based on Eq. (160). See the legend of Fig. 17a.

unknowns. One would have to try to evaluate x and y at concentration c by Monte Carlo methods or by other numerical methods (94). It appears one can actually use Eqs. (165), (168), and (169) to obtain one equation in one unknown ($\hat{B}M_1$), but the procedure for doing it is complicated and must be tested thoroughly.

Other Types of Indefinite Self-Associations

Suppose no trimer, pentamer, heptamer, etc. were present. Two types of self-associations arise: one in which all molar equilibrium constants are equal, and one for which K_{12} differs from all other molar equilibrium constants. If all molar equilibrium constants are equal then one obtains (93)

$$\begin{aligned} C &= C_1 + 2kC_1{}^2 + 4k^3C_1{}^4 + 6k^5C_1{}^6 + \cdots \\ &= C_1[1 + 2kC_1\{1 + 2k^2C_1{}^2 + 3k^4C_1{}^4 + \cdots\}] \\ &= C_1\left[1 + \frac{2kC_1}{(1 - k^2C_1{}^2)^2}\right], \qquad \text{if } kC_1 < 1 \end{aligned} \qquad (170)$$

Similarly, one notes that ξ (see Eq. 134) becomes

$$\xi = \frac{2\left[1 + \dfrac{kC_1}{(1 - k^2C_1{}^2)}\right]}{\left[1 + \dfrac{2kC_1}{(1 - k^2C_1{}^2)^2}\right]} - \frac{\left[1 + \dfrac{2kC_1}{(1 - k^2C_1{}^2)^2}\right]}{\left[1 + \dfrac{4kC_1(1 + k^2C_1{}^2)}{(1 - k^2C_1{}^2)^3}\right]} \tag{171}$$

This equation has only one unknown, kC_1. Since $0 \leq kC_1 \leq 1$, one solves for kC_1 by successive approximations. Once kC_1 is known, it is an easy matter to obtain C_1 from Eq. (170). Now let $X = kC_1$; a plot of X vs C_1 will have a slope of k. Here $k = (1000K/M_1)$ is the intrinsic equilibrium constant.

Now suppose that $K_{12} \neq K_{24}$, K_{26}, etc., but that $K_{24} = K_{26} = \cdots = K$. Letting $k_{12} = 1000K_{12}/M_1$ and $k = 1000K/M_1$, the expression for the total concentration of the associating solute is (95)

$$C = C_1\left[1 + \frac{2k_{12}C_1}{(1 - kk_{12}C_1{}^2)^2}\right]$$
$$= C_1\left[1 + \frac{2x}{(1 - xy)^2}\right] \tag{172}$$

Here $y = kC_1$ and $x = k_{12}C_1$. Note that $0 \leq x \leq 1$ and $0 \leq y \leq 1$. For this association the expressions for M_1/M_{nc} and M_1/M_{wc} become

$$\frac{M_1}{M_{nc}} = \frac{\left[1 + \dfrac{x}{(1 - xy)}\right]}{\left[1 + \dfrac{2x}{(1 - xy)^2}\right]} \tag{173}$$

and

$$\frac{M_1}{M_{wc}} = \frac{\left[1 + \dfrac{2x}{(1 - xy)^2}\right]}{\left[1 + \dfrac{4x(1 + xy)}{(1 - xy)^3}\right]} \tag{174}$$

The expression for ξ (see Eq. 134) would have two unknowns x and y, which would have to be solved at every point (at every value of ξ that was used). We would have to use methods similar to those used with the sequential indefinite self-association for which $k_{12} \neq k$. Some recent developments in our laboratory indicate that we can use Eqs. (172) and (173) to obtain an equation in one unknown, $\hat{B}M_1$. We are currently evaluating and testing this idea.

Indefinite self-associations may be important in the self-assembly of virus coat protein subunits. One could have a linear or helical association of subunits, and equations that are similar to those used here have been developed for these situations. For more details the reader should consult the papers by Chun (96). Pekar and Frank (97) have studied the self-association of insulin near neutral pH; at pH 7.4 they believe that the self-association can be described by

$$nP_1 \rightleftarrows qP_2 + hP_6 + jP_{12} + mP_{18} + \cdots \tag{175}$$

In other words, the higher aggregates are multiples of the hexamer. For bovine insulin A the monomer molecular weight is $M_1 = 5733$ (98).

The Monomer-Dimer-Trimer and Related Self-Associations

A monomer–dimer–trimer association is described by (29, 30)

$$nP_1 \rightleftarrows qP_2 + mP_3 \tag{176}$$

When Eq. (119) applies, the following relations obtain.

$$c = c_1 + K_2 c_1^2 + K_3 c_1^3 \tag{177}$$

$$\frac{cM_1}{M_{na}} = c_1 + \frac{K_2 c_1^2}{2} + \frac{K_3 c_1^3}{3} + \frac{BM_1 c^2}{2} \tag{178}$$

$$\frac{1}{\dfrac{M_1}{cM_{wa}} - BM_1} = \frac{cM_{wc}}{M_1} = c_1 + 2K_2 c_1^2 + 3K_3 c_1^3 \tag{179}$$

and

$$\ln f_a = \ln f_1 + BM_1 c \tag{180}$$

or

$$\alpha = cf_a = c_1 \exp(BM_1 c) \tag{181}$$

We can combine Eqs. (177–179) and (181) to obtain (29, 30)

$$\frac{6cM_1}{M_{na}} - 5c = 2c_1 + 3BM_1 c^2 - \frac{1}{\dfrac{M_1}{cM_{wa}} - BM_1}$$

$$= 2\alpha \exp(-BM_1 c) + 3BM_1 c^2 - \frac{1}{\dfrac{M_1}{cM_{wa}} - BM_1} \tag{182}$$

This equation contains only one unknown, BM_1, which can be solved for by successive approximations. Alternatively Eq. (182) can be recast as

$$\frac{6M_1}{M_{na}} - 5 = 2f_a \exp(-BM_1 c) + 3BM_1 c - \frac{1}{\dfrac{M_1}{M_{wa}} - BM_1 c} \tag{183}$$

The only unknown here is BM_1. This can be solved for by successive approximation. Instead of solving Eq. (182) or (183) point by point, one could set up an array of equations from several (20 to 30 or more) data points, and use a Monte Carlo procedure to find the best BM_1 as measured by the sum of the square of the error or by the sum of the absolute value of the error (99). Here the error ε would be defined as

$$\frac{\Delta\left(\dfrac{6M_1}{M_{na}} - 5\right)}{\left(\dfrac{6M_1}{M_{na}} - 5\right)_{obs}} = \varepsilon \tag{184}$$

and

$$\Delta\left(\frac{6M_1}{M_{na}} - 5\right) = \left(\frac{6M_1}{M_{na}} - 5\right)_{calc} - \left(\frac{6M_1}{M_{na}} - 5\right)_{obs} \tag{184a}$$

Sometimes with very strong self-associations, the plot of $[(M_1/M_{wa}) - 1]/c$ vs c, which is required for the evaluation of $\ln f_a$ (see Eq. 129), may be quite steep in the low concentration region. Figure 18 shows such a plot for a simulated monomer–dimer–trimer association; the intercept at $c = 0$ is $-K_2 + BM_1$. The greatest contribution to the integral required for $\ln f_a$,

$$\ln f_a = \int_0^c \left(\frac{M_1}{M_{wa}} - 1\right) \frac{dc}{c}$$

comes from the region of lowest solute concentration. This is the area where the experimental error may be the worst; and this may cause problems in the application of Eqs. (183) and (184). One way to avoid this is to calculate $\ln f_a/f_{a*}$, where

$$\ln f_a/f_{a*} = \int_{c*}^c \left(\frac{M_1}{M_{wa}} - 1\right) \frac{dc}{c} \tag{185}$$

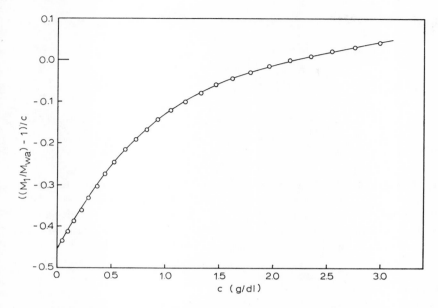

FIG. 18. Plot required for the evaluation of $\ln f_a$ (see Eq. 129). Here a simulated monomer–dimer–trimer association was used; $K_2 = 0.65$ dl/g, $K_3 = 0.5$ dl/g, and $BM_1 = 0.2$ dl/g. The intercept of this plot is $-K_2 + BM_1 = -0.45$. The actual shape of this curve depends on the values of the K_i and the BM_1; with stronger self-associations this plot becomes much steeper in the vicinity of $c = 0$.

Here c^* is a low concentration; the choice is arbitrary. What we want to do is get rid of the troublesome part of the $[(M_1/M_{wa}) - 1]/c$ vs c plot. We can then recast Eq. (183) as

$$\frac{f_1}{f_{1*}} = \frac{\left[\dfrac{6M_1}{M_{na}} - 5 + 3BM_1 c - \dfrac{1}{\dfrac{M_1}{M_{wa}} - BM_1 c}\right]}{\left[\dfrac{6M_1}{M_{na*}} - 5 + 3BM_1 c^* - \dfrac{1}{\dfrac{M_1}{M_{wa*}} - BM_1 c^*}\right]} \qquad (186)$$

Multiplication of both sides of this equation by $\exp[BM_1(c - c^*)]$ leads to (99)

$$\frac{f_a}{f_{a*}} = \frac{f_1}{f_{1*}} \exp[BM_1(c - c^*)]$$

$$= \frac{\left[\dfrac{6M_1}{M_{na}} - 5 + 3BM_1c - \dfrac{1}{\dfrac{M_1}{M_{wa}} - BM_1c}\right] \exp[BM_1(c - c^*)]}{\left[\dfrac{6M_1}{M_{na*}} - 5 + 3BM_1c^* - \dfrac{1}{\dfrac{M_1}{M_{wa*}} - BM_1c^*}\right]} \qquad (187)$$

Here one can set up an array of data points and use a Monte Carlo procedure to find the best value of BM_1 to fit the array. We did this with the self-association of adenosine-5'-triphosphate (ATP) in isotonic saline, and showed that a monomer–dimer–trimer self-association gave the best description of the observed self-association. Figure 19 shows a plot of

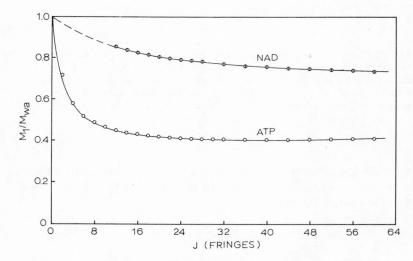

FIG. 19. Self-association of small, ionizable molecules. These plots of M_1/M_{wa} vs J give a comparison of the self-association of two nucleotides, Adenosine-5'-triphosphate (ATP) and Nicotine adenine dinucleotide (NAD), in isotonic saline at 20 °C. These solutions were dialyzed in a hollow fiber dialyzer with a 200 Dalton cutoff. It is evident that ATP associates more strongly than the NAD. ATP undergoes a monomer-dimer-trimer association (99). Experiments are still underway with NAD; preliminary results suggest a sequential, indefinite self-association may be present.

M_1/M_{wa} vs c at $10\,°C$ for ATP in isotonic saline; on the same plot we have also shown some preliminary results with the self-association of NAD (nicotine adenine dinucleotide) in isotonic saline. The self-association for ATP (99) is stronger than that for NAD (100), purine and cytidine (33, 73), or that observed with various nucleosides (74, 75).

Equations analogous to Eq. (182) or (183) can be developed for other related self-associations, such as a monomer–dimer–tetramer. For this association, the analog of Eq. (183) is

$$\frac{8M_1}{M_{na}} - 6 = 3c_1 + 4BM_1c - \frac{1}{\dfrac{M_1}{M_{wa}} - BM_1c} \tag{188}$$

Other Self-Associations

Teller (101) has considered the case of discrete self-associations in which the equilibrium constants are equal. Šolc and Elias (85) have given a very detailed treatment of a self-association involving a heterogeneous unimer (monomer). The discussion in this paper so far has been restricted to self-associations involving a homogeneous unimer (or monomer). Šolc and Elias (85) have given a very elegant treatment of a more complicated case. Their paper is the opening wedge in a vast area.

Factors Influencing Self-Associations

With a self-associating solute one can only add the solute itself to the solution; the subsequent self-association equilibrium that sets in depends on various factors such as temperature, pH, ionic strength, and the presence or absence of other additives or ions. Various types of chemical interactions may be involved. For example, with soaps and detergents in aqueous solution, hydrophobic interactions are involved. These associations can be influenced by the position of the polar group for polar detergents as well as by the ionic strength of the medium (69–72). With purine, cytidine, and various nucleosides, it is believed that base stacking is an important factor (33, 73–75). Ts'o (75) was able to prepare a chemically modified nucleoside in which no hydrogen bonds could be formed, yet it associated very strongly in aqueous solution. This association was attributed to base stacking (75). Recently we have done some studies on the temperature-dependent self-association of the disodium salt of adenosine-5'-triphosphate (ATP) dissolved in and dialyzed against isotonic saline (0.154 M NaCl); the availability of a hollow fiber dialyzer with a low molecular weight cutoff made these studies possible (99). Our results indicated that ATP underwent a monomer–dimer–trimer self-association; the association was greater at lower temperatures. Furthermore the molar association equilibrium constant for the dimerization was greater than that observed by others for purine, cytidine, or various nucleosides. One would think that the triphosphate group, since it ionizes, would inhibit self-association; yet our results indicated a much stronger self-association.

With proteins metal ions are sometimes involved. Kakiuchi (102) showed that Zn^{2+} is needed for the self-association of an anylase obtained from *B. subtilis*. Electrostatic repulsions and attractions are involved in the self-association of the β-lactoglobulins at low pH (pH 2 to 3). An increase of salt increases the degree of aggregation; Types A, B, and C show a monomer–dimer association under these conditions (31, 35, 91, 103–105). In the pH range 4 to 5.7 the association behavior is quite different. β-Lactoglobulin A shows a very strong, temperature-dependent self-association which has been characterized as an indefinite (32, 93) or as a dimer–octamer (105, 106). Sedimentation equilibrium experiments on three different batches of β-lactoglobulin A in acetate buffer at pH 4.7 and ionic strengths varying from 0.1 to 0.16 have shown that the 18,422 Dalton monomer unit is present and would be the limiting species in the vicinity of zero protein concentration. Experiments carried out using a photoelectric scanner have indicated apparent weight average (M_{wa}) molecular weights as low as 21,000 at very low concentrations, and the trend of these M_{wa} would go to the 18,422 unit. The genetic variants β-lactoglobulins B and C do not associate as strongly. McKenzie (107) maintains that —COOH groups are involved in the strong self-associations of β-lactoglobulin A in the pH range 4 to 5.7; he diminished the self-association by chemically modifying the —COOH group. The self-association of three genetic variants of α_{s_1}-casein were studied by Schmidt (108); two of the variants showed similar behavior, whereas the other one differed. The effect of various ions, presumably by ion binding, on the self-association of apoferritin has been shown by light-scattering experiments carried out on the apoferritin in different buffer solutions (109). Chymotrypsinogen associates only at low ionic strengths (110, 111) whereas chymotrypsin associates at higher ionic strength (80). The addition of diisopropyl fluorophosphate (DIP) to chymotrypsin stops the enzymatic activity, but it does not stop the self-association (38). Eisenberg and his associates (112, 113) have shown that bovine lactate dehydrogenase associates; Chun et al.

(*114*) have shown that this association is a sequential indefinite self-association. It was shown by Eisenberg and his associates (*113*) that a small amount of diethyl stilbesterol inhibited the self-association; this indicates the presence of hydrophobic bonds. Furthermore they were able to cross-link the associating species and still show some biological activity which indicated that the association and active sites were in different locations.

Clearly from this discussion it is apparent that various forces are involved in self-association, and that genetic variation can also influence the self-association. A very interesting discussion about the forces involved in protein associations and methods to test for them is given in the paper by Timasheff (*115*).

Other Methods for Studying Self-Associations

Equilibrium Thermodynamic Methods

Since M_{na} and M_{wa} are interrelated for self-associating solutes, one can use any technique that will give values of M_{wa} or M_{na} and perform a series of experiments with several solutions of different concentrations. Then plots of M_{na} or M_{wa} vs c or M_1/M_{na} or M_1/M_{wa} vs c can be constructed, and the analysis can be done in the same way that has been described here. For instance, M_1/M_{wa} is obtained from plots of M_1/M_{na} vs c since (see Eq. 132) (*90*)

$$\frac{M_1}{M_{wa}} = \frac{d}{dc}\left(\frac{cM_1}{M_{na}}\right) = \frac{M_1}{M_{na}} + c\frac{d}{dc}(M_1/M_{na}) \tag{132}$$

Similarly one notes that (*90*)

$$\ln f_a = \int_0^c \left(\frac{M_1}{M_{na}} - 1\right)\frac{dc}{c} + \left(\frac{M_1}{M_{na}} - 1\right) \tag{133}$$

The big disadvantage here is that each point on the plots of M_{wa} or M_{na} vs c corresponds to one solution, so a large amount of material is required. On the other hand one can cover the same range of M_{na} or M_{wa} vs c in sedimentation equilibrium experiments with only a few (four or more) solutions of different initial concentrations.

Two techniques that give M_{wa} are elastic light scattering and low-angle X-ray scattering. Extensive studies on the self-association of the β-lacto-globulins A and B, using light scattering, have been reported by Timasheff and Townend and their associates (*103, 106*). At low pH (pH 2 to 3.5) the

β-lactoglobulins A and B undergo a monomer–dimer association. These associations have also been studied by sedimentation equilibrium experiments, and good agreement has been found between the two techniques (*35, 91*). M_{na} can be obtained from high-speed membrane osmometry or vapor pressure osmometry. The monomer–dimer self-association of soybean proteinase inhibitor was studied by Harry and Steiner (*116*) using high-speed membrane osmometry. The self-association of purine in aqueous solutions has been examined by vapor pressure osmometry (*75, 117*) and by sedimentation equilibrium (*33, 73*). Excellent agreement has been obtained by the two methods; both methods indicated the presence of a sequential indefinite self-association. This association is attributed to base stacking. Elias (*72*) has followed the self-association of some non-ionic detergents by light-scattering, vapor pressure osmometry, and sedimentation equilibrium experiments; the degree of aggregation and the equilibrium constants obtained by the three methods agreed remarkably well.

Transport Methods

Gilbert and his associates (*118–121*) have proposed a method for analyzing self-associations from the shape of the moving boundary in a sedimentation velocity experiment. In order to solve the continuity equation when self-associations occur, Gilbert (*118*) was forced to make four assumptions: (a) The centrifugal field is constant, (b) the cell has a constant cross-sectional area, (c) there is no diffusion in a moving boundary, and (d) the velocity of the n-mer relative to the monomer is constant. For the sedimentation velocity experiment the first three assumptions are false; only the last assumption may be true. The neglect of diffusion was necessary so that the continuity equation could be solved. Only recently with very sophisticated computational procedures has it been possible to include diffusion in the continuity equation for self-associations (*122, 123*) or mixed associations (*124, 125*). Nevertheless, Gilbert's methods did stir up quite a bit of interest in the study of self-associations. His theory indicated the moving boundary should be bimodal for a monomer–n-mer association when $n \geq 3$. However, it has been shown by Fujita (*126*) that one may still encounter unimodal boundaries for a monomer–trimer association under some conditions. Cox, who has done some elegant computer simulation studies on self-associations, has shown that unimodal boundaries can be encountered with monomer–trimer and monomer–tetramer associations (*122, 123*). For a monomer–trimer association which has $M_1 \leq 50,000$ Daltons, he has shown that diffusion can mask the bimodality; the same thing can happen in a monomer–tetramer association if $M_1 \leq 20{,}000$. It has been pointed out that the presence of dimer in these associations could cause unimodal peaks (122, 123). The Gilbert method may fail when intermediate species coexist in rapid equilibrium with a monomer and its highest n-mer (127). Although the molecular weights of the associating species have the same relation $M_j = jM_1$ ($j = 2, 3, \ldots$), there is no known relation between sedimentation coefficients of the associating species. If we knew the shape of the species we might be able to predict the relation between the sedimentation coefficients of the associating species. Finally, at present no plots comparable to those based on Eqs. (134)–(136) are available for sedimentation velocity experiments; clearly, the sedimentation equilibrium method does have an advantage.

The Gilbert method has also been applied to moving boundary electrophoresis (128). The application of the Gilbert method to various transport methods has been discussed by Cann (125) and also by Nichol, Bethune, Kegeles, and Hess (128). Winzor and Scheraga (38) have tested the Gilbert method using gel filtration chromatography. They have shown that derivatives of the elution profiles resemble the schlieren patterns obtained by electrophoresis or sedimentation velocity. In fact, they claim and do demonstrate a difference in the derivatives of the elution profiles between self-associating and noninteracting proteins. From these elution profiles they can calculate the elution volume, V_e, which for self-associations exhibits a concentration dependence similar to that exhibited by sedimentation coefficients of self-associating solutes. With self-associations they measure an apparent weight-average elution volume, $V_{ew\ app}$. The concentration dependence of the elution volume is quite different for noninteracting and for self-associating systems. In fact a plot of $V_{ew\ app}$ vs c resembles a plot of $1/s_{wa}$ vs c; here s_{wa} is the apparent weight-average sedimentation coefficient. The elution volume, V_e, is proportional to molecular weight for a series of polymers (129), and this fact can be used to estimate molecular weights from gel filtration chromatography. This method has the advantage of speed and simplicity; and it has also been applied to a study of the mixed association between lysozyme and ovalbumin (130). More details about this technique will be found in the monograph by Winzor and Nichol (37).

A very beautiful and elegant method for studying chromatography directly on a column, and especially the chromatographic behavior of chemically reacting systems, is the scanning method developed by Brumbaugh and Ackers (130, 131). Quartz columns packed with sephadex are used. A high intensity lamp is placed on one side of the column and a photomultiplier is placed on the other side; the output from the photomultiplier is fed to a computer. A motor is used to move the column up or down so that various levels can be scanned. Even though light scattering causes a high background absorbance with only buffer and the sephadex gel, Brumbaugh and Ackers (130, 131) were able to show a linear relation between protein concentration and absorbance in the absorbance range of 2 to 3. With these experiments the weight-average partition coefficients, σ_w, between the gel phase and the liquid phase can be measured. The weight-average elution volume, V_{ew}, can also be obtained. An empirical relation between partition coefficients and molecular weights has been established (131). For self-associations both σ_w and V_{ew} are functions of the total concentration of the associating solute, and they can be used to estimate the equilibrium constant (3). Henn and Ackers (132) have done very elegant studies of the monomer–dimer self-association of D-amino acid oxidase apoenzyme; this association was studied at several temperatures. The van't Hoff plot of ln K_2 vs $1/T$ gave a reversed S-shape curve. This was interpreted to mean that a conformational equilibrium accompanied the self-association. With the aid of other physical methods, optical rotatory dispersion and concentration difference spectra in the UV region, they were able to calculate the conformational equilibrium constant.

Chun et al. (114) have shown that one can obtain the weight fraction of monomer, f_1, from values of the weight-average partition coefficients, σ_n, when self-associations are present. They have developed equations applicable to various types of self-associations. The σ_w could also be related empirically to molecular size. Chun et al. (114) have shown that the self-association of bovine liver L-glutamate dehydrogenase reported by Eisenberg and Tompkins (112) was a sequential indefinite self-association. Chun et al. (114) also studied this self-association by analytical gel chromatography; their plot of σ_w vs c agreed with the plot predicted for a linear indefinite self-association.

Godschalk (133) has developed a very sophisticated, computational method for determining the translational diffusion coefficients of self-associating species. The method works best with materials having an absorption spectrum in the range of 220 to 560 nm, so that a photoelectric scanner equipped with a data acquisition system can be used to reduce the tediousness of the calculations. With this method it is possible to evaluate diffusion coefficients and also to enumerate the number of associating species, if this is not known a priori.

In the discussion of the other methods for studying self-associations, we have not considered what would happen if both thermodynamic (equilibrium) and transport experiments on the same associating solute were carried out under identical solution conditions. Would it be possible to evaluate additional information about the associating solute, such as the sedimentation coefficients of the associating species? Such experiments have been performed. Kakiuchi and Williams (134) have studied the self-association of a γ-G globulin from multiple myeloma; the solvent was 8 M urea, buffered at pH 7.

If we assume that interacting flows are absent in a multicomponent system, then we can combine the two techniques. This is an assumption that one is forced to make at present. The apparent weight-average sedimentation coefficient, s_{wa}, is defined by (42)

$$s_{wa} = \frac{s_{wc}}{1 + gc} \quad \text{(Model I)} \tag{189}$$

or by a second model (42)

$$\frac{1}{s_{wa}} = \frac{1}{s_{wc}} + gc \quad \text{(Model II)} \tag{190}$$

Here s_{wc}, the weight-average sedimentation coefficient, is defined by

$$s_{wc} = \frac{\sum c_i s_i}{c} \tag{191}$$

For a monomer–dimer association

$$s_{wc} = \frac{s_1 + K_2 c_1 s_2}{1 + K_2 c_1} \tag{192}$$

The quantity g or $\boldsymbol{g}$ is the hydrodynamic concentration dependence parameter associated with ordinary sedimentation coefficient measurements on nonassociating solutes. Note that s_1 is the sedimentation coefficient of the monomer at infinite dilution, and s_2 is the sedimentation coefficient of the dimer. Thus

$$\lim_{c \to 0} s_{wa} = s_1 \tag{193}$$

or

$$\lim_{c \to 0} 1/s_{wa} = 1/s_1 \tag{193a}$$

For very strong self-associations it may be difficult to obtain s_1 from the limiting slope of a plot of s_{wa} or $1/s_{wa}$ vs c; thus one may have to estimate s_1 from experiments conducted under nonassociating conditions. It will have to be assumed that s_1 does not change with the different solution conditions. Kakiuchi and Williams (111, 134) pointed out that Eq. (189) became linear at high solute concentrations so that

$$\frac{1}{s_{wa}} \simeq \frac{1}{s_2}(1 + gc) \qquad (194)$$

If the slope of a plot of $1/s_{wa}$ vs c is taken at a very high concentration, one obtains g/s_2 as the slope and $1/s_2$ as the intercept of the tangent line at zero concentration. Clearly this method depends on where one draws the slope. There are ways to overcome this limitation; we will illustrate it with the second model (see Eq. 190).

Equation (190) suggests that a plot of $1/s_{wa}$ vs c behaves somewhat in a manner described by Fig. 20. The decrease in $1/s_{wa}$ is due to the self-association; the quantity gc increases linearly with c, so that the combination of the two factors causes a minimum in the plot of $1/s_{wa}$ vs c. This behavior is similar to that encountered with plots of M_1/M_{wa} vs c for nonideal self-associations. The actual shape of plots of $1/s_{wa}$ vs c depends

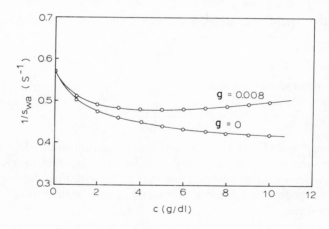

FIG. 20. Sedimentation coefficients of self-associating solutes. Here we used a simulated monomer–dimer association with $s_1 = 1.75s$, $s_2 = 2.85s$ and $K_2 = 0.35$ l/g. The lower curve shows a plot of $1/s_{wa}$ vs c for which there is no hydro-dynamic concentration dependence ($g = 0$). Model II (Eq. 190) for describing s_{wa} was used here. The upper curve shows the effect of the hydrodynamic concentration dependence parameter ($g = 0.008$ l/g). Note the resemblance of these curves to plots of M_1/M_{wa} for self-associating solutes.

on the values of K_2, s_1, s_2, and g. The similarity of the plot of $1/s_{wa}$ vs c and M_1/M_{wa} vs c suggests that we try to develop equations that eliminate the gc term, so that the resulting equation only contains one or more unknowns (s_2, etc.). One way to eliminate g is to note that

$$\frac{1}{s_{wa}} - \frac{2}{c}\int_0^c \frac{dc}{s_{wa}} = \frac{1}{s_{wc}} - \frac{2}{c}\int_0^c \frac{dc}{s_{wc}} \tag{195}$$

For a monomer–dimer association, Eq. (195) becomes

$$\frac{1}{s_{wa}} - \frac{2}{c}\int_0^c \frac{dc}{s_{wa}} = \frac{1 + K_2 c_1}{s_1 + K_2 c_1 s_2} + \frac{2}{c}\int_0^c \left(\frac{1 + K_2 c_1}{s_1 + K_2 c_1 s_2}\right) dc \tag{196}$$

Since c_1, K_2, and s_1 are known, there is only one unknown, s_2, which is evaluated by successive approximations. The easiest way to evaluate $\int_0^c (dc/s_{wc})$ is to use a computer; for each choice of s_2, the integral has to be evaluated. Another way to estimate s_2 is to use the equation

$$\frac{1}{s_{wa}} - c\left[\frac{d(1/s_{wa})}{dc}\right] = \frac{1}{s_{wc}} - c\left[\frac{d(1/s_{wc})}{dc}\right] \tag{197}$$

This equation contains only one unknown, s_2, which can be evaluated by successive approximations. Once s_2 is known then it is possible to calculate g from the relation

$$\frac{1}{s_{wa}} - \frac{1}{s_{wc}} = gc \tag{198}$$

since a plot of $[(1/s_{wa}) - (1/s_{wc})]$ vs c has a slope of g. The analysis could also be done with the other model (Eq. 189) to describe s_{wa}, and the analysis can be extended to other self-associations, including a sequential, indefinite self-association. A test of these methods has been applied to the self-association of the γ-G globulin studied by Kakiuchi and Williams (134); the results are listed in Table 3. Note that g and g have different values. Also note that s_2 evaluated using either model, Eq. (189) or (190), to described s_{wa} agreed quite well with each other. The disagreement with the Kakiuchi and Williams method may reflect the fact that the value of s_2 depends on how and where one takes the tangent to the curve of $1/s_{wa}$ vs c.

It is also possible to evaluate s_{za}, the apparent z-average sedimentation coefficient (42, 135), since with Model II (Eq. 190), s_{za} is given by

$$s_{za} = \frac{d(cs_{wa})}{dc} = s_{wa} + c\frac{d(s_{wa})}{dc} \tag{199}$$

TABLE 3
Estimation of s_2 from the Data of Kakiuchi and Williams $(134)^a$

Model	s_2 (sec)	g (dl/g)	g (dl/g)
I (Eq. 189)	9.4^b	0.32	—
I (Eq. 189)	10.3 ± 0.20^c	0.35 ± 0.03	—
II (Eq. 190)	10.25 ± 0.20^d	—	0.034 ± 0.0002

$^a s_1 = 7.25$; $K_2 = 21.5$ dl/g.
^{b}Values reported by Kakiuchi and Williams (134) using a method based on Eq. (194).
^{c}Method based on Eq. (35) of Weirich, Adams, and Barlow (42); nine data points were used for estimating s_2.
^{d}Method based on Eq. (197); see also Eq. (24) of Weirich, Adams, and Barlow. Twelve data points were used for the estimation of s_2.

One can also get s_{zc} or s_{za} with Model I. In fact, if one can measure any weight-average property, X_{wc}, of a self-associating solute, then it is possible to obtain the z-average property (42), X_{zc}, since

$$X_{zc} = \frac{d(cX_{wc})}{dc} = X_{wc} + c\frac{d(X_{wc})}{dc} \tag{200}$$

Here X_{wc} could be a weight-average partition coefficient, σ_{wc}, elution volume, $V_{e\,wc}$, or any weight or apparent weight-average property that can be measured. It is also possible to obtain a number-average property, X_{nc}, from a combination of equilibrium and transport methods (42). It should be evident that one can combine equations for X_{wc} or its apparent value, X_{wa}, in a manner similar to that done with s_{wc} or s_{wa} so that X_1, X_2, and other quantities, such as a concentration dependence parameter, can be evaluated. By combining equilibrium and transport techniques we can learn far more about the self-associating solute than we could learn from either technique alone. This is a very new development, and more work of this kind should appear in the future.

Acknowledgments

This work was supported by grants from the Robert A. Welch Foundation (A485) and from the National Science Foundation (GB 32242A1). Will E. Ferguson is a Robert A. Welch Foundation Postdoctoral Fellow.

REFERENCES

1. T. Svedberg and H. Rinde, *J. Amer. Chem. Soc.*, 46, 2677 (1924).
2. J. D. Watson, *The Molecular Biology of the Gene*, 2nd ed., Benjamin, New York, 1970, p. 61.

3. T. Svedberg and K. O. Pedersen, *The Ultracentrifuge*, Clarendon, Oxford, 1940.
4. J. W. Williams, K. E. Van Holde, R. L. Baldwin, and H. Fujita, *Chem. Revs.*, *58*, 715 (1958).
5. H. Fujita, *Mathematical Theory of Sedimentation Analysis*, Academic, New York, 1960.
6. H. Rinde, Ph.D. Dissertation, University of Upsala, 1928.
7. J. S. Johnson, K. A. Kraus, and G. Scatchard, *J. Phys. Chem.*, *64*, 1867 (1960).
8. F. E. LaBar and R. L. Baldwin, *Ibid.*, *66*, 1952 (1962).
9. J. S. Johnson, K. A. Kraus, and G. Scatchard, *Ibid.*, *58*, 1034 (1954).
10. J. S. Johnson and K. A. Kraus, *Ibid.*, *63*, 440 (1959).
11. J. S. Johnson, K. A. Kraus, and R. W. Holmberg, *J. Amer. Chem. Soc.*, *78*, 26 (1956).
12. F. C. Hentz and J. S. Johnson, *Inorg. Chem.*, *5*, 1337 (1966).
13. J. S. Johnson, G. Scatchard, and K. A. Kraus, *J. Phys. Chem.*, *63*, 787 (1959).
14. J. Brandrup and E. H. Immergut, eds., *Polymer Handbook*, Wiley-Interscience, New York, 1966.
15. D. A. Albright and J. W. Williams, *J. Phys. Chem.*, *71*, 2780 (1967).
16. T. H. Donnelly, *Ibid.*, *70*, 1862 (1966).
17. T. H. Donnelly, *Ann. N. Y. Acad. Sci.*, *164*, 147 (1969).
18. Th. G. Scholte, *J. Polym. Sci.*, *A-2*, *6*, 111 (1968).
19. Th. G. Scholte, *Ann. N. Y. Acad. Sci.*, *164*, 156 (1969).
20. Th. G. Scholte, *Eur. Polym. J.*, *6*, 51 (1970).
21. S. W. Provencher, *J. Chem. Phys.*, *46*, 3229 (1967).
22. L.-O. Sundelöf, *Ark. Kemi*, *29*, 297 (1968).
23. M. Gehatia and D. R. Wiff, *Advan. Chem. Ser.*, *125*, 216 (1973).
24. E. T. Adams, Jr., P. J. Wan, D. A. Soucek, and G. H. Barlow, *Ibid.*, *125*, 235 (1973).
25. J. Vinograd and J. E. Hearst, *Fortschr. Chem. Org. Naturstoffe*, *20*, 372 (1962).
26. M. Meselson and F. W. Stahl, *Proc. Natl. Acad. Sci. U. S.*, *44*, 671 (1958).
27. S. Kaufman and A. J. Fryar, Paper presented at the 160th National Meeting of the American Chemical Society, Chicago, September 14–18, 1970. Abstract No. Coll 88.
28. J. J. Hermans, *Ann. N. Y. Acad. Sci.*, *164*, 122 (1969).
29. E. T. Adams, Jr., *Biochemistry*, *4*, 1646 (1965).
30. E. T. Adams, Jr., *Fractions*, No. 3, 1967.
31. D. A. Albright and J. W. Williams, *Biochemistry*, *7*, 67 (1968).
32. E. T. Adams, Jr. and M. S. Lewis, *Ibid.*, *7*, 1044 (1968).
33. K. E. Van Holde, G. P. Rossetti, and R. D. Dyson, *Ann. N. Y. Acad. Sci.*, *164*, 279 (1969).
34. P. W. Chun, S. J. Kim, J. D. Williams, W. T. Cope, L. H. Tang, and E. T. Adams, Jr., *Biopolymers*, *11*, 197 (1972).
35. L. H. Tang and E. T. Adams, Jr., *Arch. Biochem. Biophys.*, *157*, 520 (1973).
36. L. W. Nichol, J. L. Bethune, G. Kegeles, and E. L. Hess, in *The Proteins*, Vol. II, 2nd ed. (H. Neurath, ed.), Academic, New York, 1964, p. 305.
37. L. W. Nichol and D. J. Winzor, *Migration of Interacting Systems*, Clarendon, Oxford, 1972.
38. D. J. Winzor and H. A. Scheraga, *Biochemistry*, *2*, 1263 (1963); *J. Phys. Chem.*, *68*, 338 (1964).
39. G. K. Ackers, *Advan. Protein Chem.*, *24*, 343 (1970).
40. S. W. Henn and G. K. Ackers, *Biochemistry*, *8*, 3829 (1969).

41. P. W. Chun, S. J. Kim, C. A. Stanley, and G. K. Ackers, *Ibid.*, *8*, 1625 (1969).
42. C. A. Weirich, E. T. Adams, Jr., and G. H. Barlow, *Biophys. Chem.*, *1*, 35 (1973).
43. P. J. Wan, Ph. D. Dissertation, Texas A & M Univ. College Station, Texas, December 1973.
44. P. J. Wan and E. T. Adams, Jr., *Polym. Preprints*, *15*, 509 (1974).
45. M. Gehatia and D. R. Wiff, *J. Polym. Sci.*, *A-2*, *8*, 2039 (1970).
46. M. Gehatia, *Polym. Preprints*, *12*, 875 (1971).
47. D. R. Wiff and M. Gehatia, *J. Macromol. Sci.—Phys.*, *B6*, 287 (1972).
48. M. Gehatia and D. R. Wiff, *Eur. Polym. J.*, *8*, 585 (1972).
49. M. Gehatia and D. R. Wiff, *J. Chem. Phys.*, *57*, 1070 (1972).
50. E. T. Adams, Jr., in *Characterization of Macromolecular Structure*, Publication 1573, National Academy of Sciences, Washington, D.C., 1968, p. 106.
51. W. D. Lansing and E. O. Kraemer, *J. Amer. Chem. Soc.*, *57*, 1369 (1935).
52. J. W. Williams, *Ultracentrifugation of Macromolecules*, Academic, New York, 1972.
53. R. L. Baldwin and K. E. Van Holde, *Fortschr. Hochpolym.-Forsch.*, *1*, 451 (1960).
54. H. W. McCormick, in *Polymer Fractionation* (M. J. R. Cantow, ed.), Academic, New York, 1967, p. 251.
55. M. Wales and S. J. Rehfeld, *J. Polym. Sci.*, *62*, 179 (1962).
56. J.-P. Merle and A. Sarko, *Macromolecules*, *5*, 132 (1972).
57. T. Bluhm and A. Sarko, *Ibid.*, *6*, 578 (1973).
58. D. J. DeRosier, P. Munk, and D. J. Cox, *Anal. Biochem.*, *50*, 139 (1972).
59. L. J. Gosting, *J. Amer. Chem. Soc.*, *74*, 1458 (1952).
60. J. W. Williams, W. M. Saunders, and J. S. Cicirelli, *J. Phys. Chem.*, *58*, 774 (1954).
61. J. W. Williams and W. M. Saunders, *Ibid.*, *58*, 854 (1954).
62. J. C. Moore, in *Characterization of Macromolecular Structure*, Publication 1573, National Academy of Sciences, Washington, D.C., 1968, p. 273.
63. M. Ezrin (ed.), *Polymer Molecular Weight Methods*, American Chemical Society Advances in Chemistry Series No. 125, 1973.
64. K. A. Granath, *J. Colloid Sci.*, *13*, 308 (1958).
65. K. A. Granath, *Makromol. Chem.*, *28*, 1 (1958).
66. J. W. A. Averink, H. Reerink, J. Boerma, and W. J. M. Jaspers, *J. Colloid Interface Sci.*, *21*, 66 (1966).
67. H. J. Cantow, *Makromol. Chem.*, *70*, 130 (1964).
68. O. Bryngdahl and S. Ljunggren, *J. Phys. Chem.*, *64*, 1264 (1960).
69. P. J. Debye, *J. Phys. Colloid Chem.*, *51*, 18 (1947); *Ann. N. Y. Acad. Sci.*, *51*, 575 (1949); *J. Phys. Colloid Chem.*, *53*, 1 (1949).
70. E. W. Anacker, R. M. Rush, and J. S. Johnson, *J. Phys. Chem.*, *68*, 81 (1964).
71. S. Ikeda and K. Kakiuchi, *J. Colloid Interface Sci.*, *23*, 134 (1967).
72. H.-G. Elias, *J. Macromol. Sci.—Chem.*, *A7*, 601 (1973).
73. K. E. Van Holde and G. P. Rossetti, *Biochemistry*, *6*, 2189 (1967).
74. T. N. Solie and J. A. Schellman, *J. Mol. Biol.*, *33*, 61 (1968).
75. P. O. P. T'so, *Ann. N. Y. Acad. Sci.*, *153*, 785 (1969).
76. H.-G. Elias and R. Bareiss, *Chimia*, *21*, 53 (1967).
77. A. Tiselius, *Z. Phys. Chem.*, *124*, 449 (1926).
78. R. F. Steiner, *Arch. Biochem. Biophys.*, *39*, 333 (1952); *44*, 120 (1953).
79. R. F. Steiner, *Ibid.*, *49*, 400 (1954).
80. M. S. N. Rao and G. Kegeles, *J. Amer. Chem. Soc.*, *80*, 5724 (1958).

81. E. T. Adams, Jr., and H. Fujita, in *Ultracentrifugal Analysis in Theory and Experiment* (J. W. Williams, ed.), Academic, New York, 1966, p. 119.

82. E. T. Adams, Jr., and J. W. Williams, *J. Amer. Chem. Soc.*, *86*, 3454 (1964).

83. M. Derechin, *Biochemistry*, *7*, 3253 (1968); *8*, 921, 927 (1969); *11*, 1120, 4153 (1972).

84. M. Derechin, Y. M. Rustrum, and E. A. Barnard, *Ibid.*, *11*, 1793 (1972).

85. K. Šolc and H.-G. Elias, *J. Polym. Sci., Polym. Phys. Ed.*, *11*, 137 (1973).

86. A. Vrij and J. Th. G. Overbeek, *J. Colloid. Sci.*, *17*, 570 (1962).

87. E. F. Casassa and H. Eisenberg, *Advan. Protein Chem.*, *19*, 287 (1964).

88. H. Edelhoch, E. Katchalski, R. H. Maybury, W. L. Hughes, Jr., and J. T. Edsall, *J. Amer. Chem. Soc.*, *75*, 5058 (1953).

89. E. Braswell, *J. Phys. Chem.*, *78*, 2477 (1968).

90. E. T. Adams, Jr., *Biochemistry*, *4*, 1655 (1965).

91. J. Visser, R. C. Deonier, E. T. Adams, Jr., and J. W. Williams, *Ibid.*, *11*, 2634 (1972).

92. P. J. Flory, *Principles of Polymer Chemistry*, Cornell Univ. Press, Ithaca, New York, 1953. Chap. 8.

93. L.-H. Tang, Ph.D. Dissertation, Illinois Institute of Technology, Chicago, December 1971.

94. F. S. Acton, *Numerical Methods That Work*, Harper and Row, New York, 1970.

95. L.-H. Tang, B. M. Escott, and E. T. Adams, Jr., Unpublished Material.

96. P. W. Chun, *Biophys. J.*, *10*, 563, 577 (1970).

97. A. H. Pekar and B. H. Frank, *Biochemistry*, *11*, 4013 (1972).

98. C. Tanford, *Physical Chemistry of Macromolecules*, Wiley, New York, 1961, p. 7.

99. W. E. Ferguson, C. M. Smith, E. T. Adams, Jr., and G. H. Barlow, *Biophys. Chem.*, *1*, 325 (1974).

100. W. E. Ferguson, B. M. Escott, J. L. Sarquis, P. J. Wan, E. T. Adams, Jr., and G. H. Barlow, Unpublished Material.

101. D. C. Teller, *Biochemistry*, *9*, 4201 (1970).

102. K. Kakiuchi, *J. Phys. Chem.*, *69*, 1829 (1965).

103. S. N. Timasheff and R. Townend, *J. Amer. Chem. Soc.*, *83*, 470 (1961).

104. J. L. Sarquis and E. T. Adams, Jr., *Arch. Biochem. Biophys.*, *163*, 442 (1974).

105. H. A. McKenzie, *Milk Proteins*, Vol. 2, Academic, New York, 1971, Chap. 14.

106. R. Townend and S. N. Timasheff, *J. Amer. Chem. Soc.*, *82*, 3168 (1960); S. N. Timasheff and R. Townend, *Ibid.*, *83*, 464 (1961); T. F. Kumosinski and S. N. Timasheff, *Ibid.*, *88*, 5635 (1966).

107. H. A. McKenzie, *Milk Proteins*, Vol. 1, Academic, New York, 1970, p. 215.

108. D. G. Schmidt, Ph.D. Dissertation, University of Utrecht, Utrecht, The Netherlands, 1969.

109. G. W. Richter and G. F. Walker, *Biochemistry*, *6*, 2871 (1967).

110. D. K. Hancock and J. W. Williams, *Ibid.*, *8*, 2598 (1969).

111. J. W. Williams, *Ultracentrifugation of Macromolecules*, Academic, New York, 1972.

112. H. Eisenberg and G. Tompkins, *J. Mol. Biol.*, *31*, 37 (1968).

113. R. Josephs, H. Eisenberg, and E. Reisler, *in Protein-Protein Interactions* (R. Jahnicke and E. Helmreich, eds.), Springer, New York, 1972, p. 57.

114. P. W. Chun, S. J. Kim, C. A. Stanley, and G. K. Ackers, *Biochemistry*, *8*, 1625 (1969).

115. S. N. Timasheff, in *Proteides of the Biological Fluids* (Proceedings of the 20th Colloquium Brugge, 1972), (H. Peeters ed.), Pergamon, Oxford, 1973, p. 511.

116. J. B. Harry and R. F. Steiner, *Biochemistry*, *8*, 5060 (1969).

117. P. O. P. T'so, I. S. Melvin, and A. C. Olson, *J. Amer. Chem. Soc.*, *85*, 1289 (1963).

118. G. A. Gilbert, *Discussions Faraday Soc.*, *20*, 68 (1955); *Proc. Roy. Soc. A*, *250*, 377 (1959).

119. G. A. Gilbert and R. C. Ll. Jenkins, *Ibid.*, *A*, *253*, 420 (1959).

120. L. M. Gilbert and G. A. Gilbert, *Nature*, *194*, 1173 (1962).

121. G. A. Gilbert, *Proc. Roy. Soc.*, *A*, *276*, 354 (1963).

122. D. J. Cox, *Arch. Biochem. Biophys.*, *142*, 514 (1971).

123. D. J. Cox, *Ibid.*, *146*, 181 (1971).

124. W. B. Goad and J. R. Cann, *Ann. N. Y. Acad. Sci.*, *164*, 172 (1969).

125. J. R. Cann, *Interacting Macromolecules*, Academic, New York, 1970.

126. H. Fujita, *Mathematical Theory of Sedimentation Analysis*, Academic, New York, 1960, pp. 209–215.

127. L. W. Nichol and H. A. McKenzie, in *Milk Proteins*, Vol. 1 (H. A. McKenzie, ed.), Academic, New York, 1970, pp. 280–288.

128. L. W. Nichol, J. L. Bethune, G. Kegeles, and E. L. Hess, in *The Proteins*, Vol. 2, 2nd ed. (H. Neurath, ed.), Academic, New York, 1964, p. 305.

129. K. Granath and P. Flodin, *Makromol. Chem.*, *48*, 160 (1961).

130. E. E. Brumbaugh and G. K. Ackers, *J. Biol. Chem.*, *243*, 6315 (1968).

131. G. K. Ackers, *Advan. Protein Chem.*, *24*, 343 (1970).

132. S. W. Henn and G. K. Ackers, *Biochemistry*, *8*, 3829 (1969).

133. W. Godschalk, *Ibid.*, *10*, 3284 (1971).

134. K. Kakiuchi and J. W. Williams, *J. Biol. Chem.*, *241*, 2781 (1966).

135. G. Kegeles, *Proc. Nat. Acad. Sci. U.S.*, *69*, 2577 (1972).

Bioseparation by Zonal Centrifuges

H. W. HSU

DEPARTMENT OF CHEMICAL AND METALLURGICAL ENGINEERING
THE UNIVERSITY OF TENNESSEE
KNOXVILLE, TENNESSEE 37916

Abstract

A number of analyses relating to prediction of instantaneous particles position in a gradient solution, gradient capacity and stability in isopycnic banding, ideal number of septa in a zonal rotor, and an optimal density gradient solution for a velocity-sedimentation, etc. has been studied. The purpose of this review is to summarize the present status in analyses and operations of zonal centrifugation and to indicate the introduction of newer techniques which may open up a large number of separation methods by zonal centrifuges.

METHODS OF CENTRIFUGATION

Three basic techniques generally used in centrifugal separations are (8, 35, 42, 43): (a) the sedimentation-velocity method, (b) the approach-to-equilibrium method, and (c) the equilibrium method. With the introduction of combinations of these methods in various ways, a large number of methods has been developed (1–10) and also a variety of terms has been used to describe these centrifuge techniques (11).

Each centrifugation method is briefly discussed in the following.

Sedimentation-Velocity Method

For this method, a centrifugal field of specified force is chosen for the system under investigation. Because the rotor is operated at high speeds, the solute particles are forced to settle at appreciable rates toward the wall of the rotor. Thus the sedimentation-velocity technique is basically a transport method; the parameter used to describe the transport of particles is called the "sedimentation coefficient." It is simply a measure of velocity of sedimentation particles per unit centrifugal force field and defined by Svedberg as (*12, 13*),

$$s = \frac{(dr/dt)}{(\omega^2 r)} = \text{Svedberg units} \equiv \sec(\times\ 10^{-13}) \tag{1}$$

The rate of particle movement is a function of molecular weight as well as particle size and shape, the difference in density between the particles and the local sustaining liquid density, the local sustaining liquid viscosity, and the centrifugal force field strength. In order to make the characteristics unique, the observed sedimentation coefficient is customarily converted to the sedimentation medium of water at $20°C$ for the particle. The conversion formula is

$$s_{H_2O,20°C} = s_{obs}\left(\frac{\mu_{G,T}}{\mu_{H_2O,20°C}}\right)\left(\frac{1 - \bar{v}\rho_{H_2O,20°C}}{1 - \bar{v}\rho_{G,T}}\right) \tag{2}$$

in which $\bar{v} = 1/\rho_p$ is the specific volume of a particle; $\mu_{G,T}$ is the viscosity of a sustaining solution at temperature T; and $\rho_{G,T}$ is the density of a sustaining solution at temperature T.

Sedimentation-Equilibrium Method (14)

Using the sedimentation-equilibrium method, the concentration distribution of particles at equilibrium positions (isopycnic banding) is usually studied. The banding solutes move outward from the center of the rotor with the velocity V_r until the solutes reach their respective locations of isodensity in the sustaining solution, then V_r vanishes and an equilibrium state is established. There are many observable properties which are characteristic of systems in a state of internal equilibrium. The effects of nonideality in any real system are so significant that the thermodynamic approach is usually used in its study.

The mathematical relationship describing sedimentation equilibrium can be obtained from equations which express the criteria of equilibrium in the system. Thus the general conditions for sedimentation equilibrium in a continuous system are:

(1) Thermal equilibrium

$$\text{Grad } T = 0 \tag{3a}$$

(2) Mechanical equilibrium

$$\text{Grad } P = 0 \tag{3b}$$

(3) Equilibrium distribution of concentration of various components

$$M_i(1 - \bar{v}_i^{(r)}\rho^{(r)})\omega^2 r = \sum_{k=1}^{N} \mu_{ik} \text{ Grad } C_k^{(r)}\big|_{T,P} \tag{3c}$$

where M_i and $\bar{v}_i$ are the molecular weight and partial specific volume of component i, ρ is the density of the system, ω is the angular velocity, r is the distance from the center of rotation, superscript (r) is for the values evaluated at this level, subscript T and P are for constant temperature and pressure, C_k is the molar volume concentration (component 0 is the principal solvent), $\mu_{ik} = (\partial\mu_i/\partial C_k)_{P,T,C_l}$ where μ_i is the chemical potential of component i and the subscript C_l signifies constancy of all molarities except that indicated in the differentiation, furthermore, $\mu_{ik} = \mu_{ki}$ due to the Maxwell relation.

It is important to note that the molecular weight of any component depends on all components in the system as expressed by the term μ_{ik}. In addition, the density ρ, which is introduced through the dependence of pressure, is clearly the density of the solution and is a function of position in the rotor. Most of the studies on the sedimentation-equilibrium method assume that $\bar{v}_i$ and ρ are independent of position in the rotor. Density changes both because of the variation in the concentration throughout the rotor and because of the increase in pressure with distance. But the dependence of the partial specific volume on concentration is often negligible (particularly for proteins). One major disadvantage in the equilibrium method is the time required to attain equilibrium. For example, in work with certain proteins of very high molecular weight, the time required to reach equilibrium might involve several weeks. And, during such long intervals, certain proteins and enzymes can be denatured.

Approach-to-Equilibrium Method (15)

The third centrifugal method, approach-to-equilibrium, preserves several advantages of the equilibrium method while eliminating the excess time consumption. This technique employs the principle that in a closed system the conditions for equilibrium are fulfilled at all times during the run at the meniscus and at the bottom of the rotor. Thus, if the intermediate stages of the sedimentation equilibrium run or the very late states of a velocity run are analyzed, molecular weight can be determined by evaluating concentration and the concentration gradient at the meniscus and the rotor bottom, although these measurements at the meniscus and the rotor bottom for a zonal centrifugation are not yet available.

When the banding solutes move outward from the axis with the velocity V_r, before solutes reach their respective zone of isodensity in the gradient, diffusion and sedimentation by mutual interference produces a pseudo-equilibrium state in a nonuniform system, where the diffusional fluxes of solutes vanish. In this pseudo-equilibrium state, there always exists a relationship between transport coefficients essentially derivable from classical thermodynamics.

The simultaneous occurrence of sedimentation and diffusion in any isothermal fluid mixture, the flux of component i, may be described as (13):

$$J_i = C_i s_i \omega^2 r - \sum_{k=1}^{N} D_{ik} \text{ Grad } C_k; \qquad i = 1, 2, 3, \ldots, N \qquad (4)$$

The set of N independent equations is nothing more than a concise statement of the linear phenomenological laws for sedimentation and diffusion obtained from the irreversible thermodynamics which includes the definitions of the sedimentation coefficient s_i and the multicomponent diffusivity D_{ik}. After a certain period of time of centrifugation, concentration gradients have been built up, and diffusion processes occur which counteract the separation by sedimentation. All irreversible processes stop when, by counterbalance of the two types of transport phenomena, the flux of solutes vanishes everywhere. This is equivalent to the attainment of sedimentation pseudo-equilibrium. Thus one has

$$J_i = 0; \qquad i = 1, 2, 3, \ldots, N \qquad (5a)$$

Therefore, the pseudo-equilibrium conditions may be written, according to Eqs. (4) and (5a),

$$C_i s_i \omega^2 r = \sum_{k=1}^{N} D_{ik} \text{ Grad } C_k; \qquad i = 1, 2, 3, \ldots, N \qquad (5b)$$

On the other hand, thermodynamics for sedimentation equilibrium given in Eq. (3c) must also hold. One defines $|\mu_{ik}|$ as the determinant of all μ_{ik} and the following determinant,

$$\Gamma_k = \begin{vmatrix} \mu_{11} \cdots & \mu_{1\ k-1} & M_1(1 - \bar{v}_1\rho) & \mu_{1\ k+1} & \mu_{1N} \\ \mu_{21} \cdots & \mu_{2\ k-1} & M_2(1 - \bar{v}_2\rho) & \mu_{2\ k+1} & \mu_{2N} \\ & \cdots & & & \cdots \\ & \cdots & & & \cdots \\ & \cdots & & & \cdots \\ & \cdots & & & \cdots \\ \mu_{N1} \cdots & \mu_{N\ k-1} & M_N(1 - \bar{v}_N\rho) & \mu_{N\ k+1} & \mu_{NN} \end{vmatrix} \qquad (6)$$

and solves the set of Eqs. (3c) with respect to Grad C_k ($k = 1, 2, 3, \ldots, N$). The expressions thus obtained are to be inserted into Eq. (5b). One then obtains

$$s_i = (C_i |\mu_{ik}|)^{-1} \sum_{k=1}^{N} \Gamma_k D_{jk}, \qquad i = 1, 2, 3, \ldots, N \qquad (7)$$

This is the general relationship between sedimentation coefficients s_i and multicomponent diffusion coefficients D_{ik} for any fluid mixture (since D_{ik} and s_i depend only on the local state variables, the conclusion, though derived from equilibrium conditions, remains valid for any state outside equilibrium).

Equation (7) shows that the sedimentation coefficient depends on the concentration and the diffusivity of the sedimenting materials. The degree of departure, however, varies markedly from one substance to another. This dependence, of itself, can provide some information about the physical properties of the macromolecules under investigation. Furthermore, the multicomponent diffusivity of the sedimenting materials, of itself, is highly concentration dependent. Because of this dependence of the sedimentation coefficient on concentration, three complications may arise:

(1) The sedimentation coefficient may not be constant throughout the experiment.

(2) The sedimentation coefficient may have to be determined at infinite dilution.

(3) The dependence of the sedimentation coefficient on concentration may cause distortion in the shape of the boundary.

These three phenomena must be taken into consideration if experimental data are to be interpreted with significance. The most serious consequence of the dependence of the sedimentation coefficient on concentration is its effect on the shape of the boundary. Diffusion will cause band broadening, which inevitably affects all systems. Another consequence from diffusion is that an initially stable band may rapidly develop instability at its lower boundary if a density inversion is created there by diffusion of a gradient material (*16, 17*). When an equilibrium band is broad, the effective density gradient is more likely to depart from linearity. Consequently, the resolution on separation is greatly reduced by this effect, unless there is a measure to calibrate the exact density gradient. In order to facilitate the calibration, one has to know all the diffusivities of the sedimenting materials in the density gradient. Those diffusivities are highly concentration-dependent, and usually none of them is available in the literature.

ZONAL CONTRIFUGATION

A zonal centrifuge rotor is basically a cylindrical pressure vessel which spins about its own axis with a means of introducing and recovering a liquid density gradient and samples either dynamically (while rotating) or statically (at rest); the mechanical structure is such that measurements of quantities such as concentration gradient and sedimentation velocity bands, which are measured by an optical method in an ultra-analytical centrifuge run, are not easily made during zonal centrifuge runs. It is thus necessary to rely on a refined theory of mathematical modeling for prediction of such quantities.

Since the first zonal centrifuge was built by Anderson (*18*) in 1954, over 50 different zonal centrifuge designs have been evaluated. These have been grouped into a series of classes according to the range of speed, the material of construction, etc. A detailed summary of classification appears elsewhere (*5*).

Loading

Dynamic Loading

To load a zonal rotor during rotation requires fluid lines connecting to both the center and the edge of the rotor. Hydrostatic pressure is equalized by leading both lines back very close to the axis of rotation. When a single seal is used the two lines are coaxial, with the line connected to the rotor center in the center of the seal. The edge line therefore comes into the seal a few millimeters from the axis, thereby creating a small difference in hydrostatic pressure between the two lines during rotation. This difference is largely equalized by the presence of fluid in the edge lines, which is denser than that found in the center line—the density difference being that between the underlay or cushion under the gradient and the overlay solution over the sample.

When loading starts, the light-end of the gradient is introduced to the edge of the rotor during rotation. Centrifugal force at this state in the loading cycle makes the rotor act as a pump which draws the gradient in. Since most gradient engines pump at a constant rate, dissolved air in the gradient tends to form air bubbles in the edge line due to the negative pressure in this line. The air bubbles cause little difficulty until the rotor is nearly full. At that time the hydrostatic pressure due to the fluid in the main rotor chamber is only partially balanced by fluid in the edge line, resulting in a high back-pressure. To solve this problem, a gradient-producing device which forms the gradient as fast as it is drawn into the rotor is required. When the gradient is in the rotor, additional dense fluid, termed the underlay or "cushion," is added to the edge until the rotor is full. Flow through the rotor is now reversed and the sample layer is introduced through the center line, displacing an equal volume of the underlay.

In all dynamically loading rotors, tapered surfaces exist which serve to funnel particle zones into the center exit line during the unloading and also to minimize mixing of the sample with the gradient and the overlay during loading. To leave the sample in contact with these slanted surfaces would result in a starting sample of uneven thickness which is too close to the axis of rotation, i.e., in a low centrifugal field. Hence a solution having a density slightly less than that of the sample layer, termed the "overlay," is added to move the sample layer out into the rotor and free of the core surface. At this juncture, flow is reversed several times and small volumes of fluid are run alternately to the center and the edge to ensure that all entrapped air has been expelled. This illustrates an additional function of the overlay and the underlay, which is to provide the fluid volume necessary to allow the sample and gradient to be moved back and forth radially in the spinning rotor.

The rotor is now accelerated to operating speed. With some rotors the seals are left attached during the entire run. It should be noted that the centrifugal force field has a very considerable stabilizing effect on a liquid density gradient.

Static Loading

Static loading is generally done through a tube connected to the bottom of the centrifuge rotor. The overlay, sample, and gradient may be introduced in that order, or the sample and overlay may be carefully added from the top after the gradient is in position. An alternate method is to fill the rotor with dense underlay and then, by withdrawing part of the underlay from the bottom, draw in the gradient, sample layer, and overlay from the top. During acceleration and reorientation these isodensity levels become paraboloids of rotation described by the equation

$$Z = r^2\omega^2/2g + Z_0 \qquad (8)$$

where r is the distance from the axis of rotation, Z is the vertical distance from the bottom of the rotor, and Z_0 is the minimum of the vertical distance of the paraboloid which is a function of the angular velocity ω (in radians per second) and can be above the bottom or below the bottom (20, 21).

In an earlier work (20) we concluded that one has to bring the rotor as slowly as possible up to a pseudo-steady-state rpm, the minimum rpm at which the isodense paraboloid will stop changing its shape with an increase of rpm; then, after this speed is reached, the rate of acceleration of the rotor does not appreciably affect the shearing forces in the liquids. Because no further variation of interfacial area occurs, dispersion due to the shearing forces in each layer disappear. However, our theoretical and experimental studies (22) later indicated that the previous conclusion was incorrect. The dispersion of sample layers in centrifugation is independent of the rate of acceleration in the startup of a rotor. The dispersion of sample layers (the resolution) depends on the configuration of a rotor and loading levels. With a given rotor at a given loading level, a loss of resolution due to dispersion from the changes of interfacial area is a constant.

During the gradient reorientation, from rest to a stable orientation in a high centrifugal force field, the shearing forces occurring in a liquid confined in a closed cylinder will cause an increase in dispersion of a reorienting gradient system. The dispersion coefficient contributed from reorientation, Δ, may be written as

$$\Delta = \frac{dS}{dt}\left[\frac{\text{area}}{\text{time}}\right] \qquad (9)$$

In a given time period, the total dispersion due to reorientation shearing forces by changing in isodensity interfacial area is

$$\sigma = \int_0^t \Delta \, dt = \int_0^t \frac{ds}{dt} \, dt = S(t) - S(0) \qquad (10)$$

From Figs. 1 and 2 the paraboloid interfacial area as a function of speeds of revolution and the loading levels of liquid for K-III and J-I rotors, one finds that an isodensity interfacial areas approach a constant value

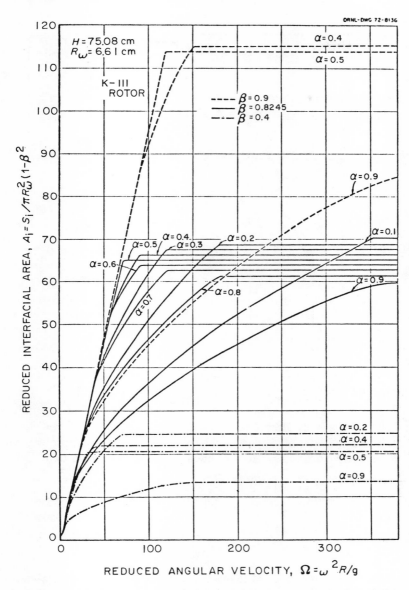

FIG. 1. Variation of reduced interfacial area with respect to speeds of revolution for a K–III rotor. α = reduced liquid loading level, β = ratio of core radius to rotor wall, H = height of rotor, and R_w = inside radius of rotor.

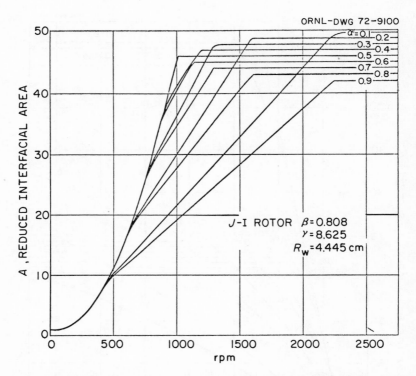

FIG. 2. Variation of reduced interfacial area with respect to speeds of revolution for a J–I rotor. γ = ratio of height to rotor wall.

(completely oriented) after a certain rotational speed for a given liquid loading level. A rotational speed, rpm = $60\omega/2\pi$, is directly proportional to an angular velocity $\omega = \sec^{-1}$. Thus Eq. (10) may be rewritten

$$\sigma = \int_0^t \frac{ds}{dt}\,dt = \int_0^{t_c} \frac{dS}{dt}\,dt + \int_{t_c}^t \frac{ds}{dt}\,dt = S(t_c) - S(0)$$

$$= \int_0^\omega \frac{ds}{d\left(\frac{1}{\omega}\right)}\,d\left(\frac{1}{\omega}\right) + \int_{\omega_c}^\omega \frac{ds}{d\left(\frac{1}{\omega}\right)}\,d\left(\frac{1}{\omega}\right) = S(\omega_c) - S(0) \qquad (11)$$

in which t_c is the time required to reach a completely reoriented paraboloid configuration, and ω_c is the angular velocity at which the paraboloid configuration is completely reoriented. From Eq. (11), one may see that $S(\omega_c)$ is a function of rotor configuration and liquid loading level only.

This finding is in agreement with our experimental investigation (22) on two-dimensional transient flow patterns and shear stress distributions, in which we have found that changing the rate of acceleration produces no noticeable effect in any flow patterns, shear stress distributions, or their magnitudes during transient periods for a given rotor.

From the foregoing analysis, we would like to make the following suggestions: One should accelerate a rotor quickly to reduce total operation time, so that a dispersion due to a molecular diffusion will be minimized. Control of rotor acceleration or deceleration is a very difficult task. Therefore, the control is unnecessary. Furthermore, at high speeds the fluctuation of a rotor speed is less than that at lower speeds and a smooth operation is easier to obtain.

We feel that the improvement of zonal centrifugation should be directed to an improvement in loading and unloading methods to reduce the dispersion from these operations. The next step will be to improve rotor configuration so that dispersion from shearing force by the reorienting gradient can be minimized.

To date the practical use of gradient reorientation has been chiefly continuous-sample-flow-with-banding centrifugation. However, it is a potentially useful method for large-scale, low-speed gradient centrifugation and also for centrifugation at very high speeds where continuously attached seals cannot be used. Use of detachable seals for high-speed rotors also presents problems because they are light in weight and have insufficient rotational momentum to allow seals to be attached without creating turbulence. As discussed below, the turnover effect can markedly decrease resolution in rotors loaded at rest. However, it does not affect gradients after a separation has been made. Simple methods for inserting the sample layer during rotation in rotors otherwise loaded and unloaded statically are required.

Gradient Runs

In order to design experimental runs, one needs to know the exact position of each particle as a function of time in a given gradient solution. A mathematical analysis for prediction of instantaneous positions and sedimentation coefficients of each particle was made by Hsu (23). His analysis is as follows:

The sedimentation of a particle which does not behave osmotically in a gradient density is a function of the following: (a) the amount and duration of the applied centrifugal force field, usually indicated by $\omega^2 t$; (b) the density and viscosity of the gradient solution medium; and (c) the size, shape, and density of the particles.

In expressing profiles of a gradient solution whose viscosity and density increase with radial distance from the axis by the polynomials

$$\mu = \mu_0(1 + \lambda_1'r + \lambda_2'r^2 + \lambda_3'r^3 + \cdots) \tag{12a}$$

$$\rho = \rho_0(1 + \varepsilon_1'r + \varepsilon_2'r^2 + \varepsilon_3'r^3 + \cdots) \tag{12b}$$

Hsu (24) has obtained mathematical expressions for (a) a particle position as a function of reduced rotation time, (b) an instantaneous particle sedimentation velocity in a given gradient solution, (c) an instantaneous observed particle sedimentation coefficient in a given gradient solution, and (d) the shear-stress exerted by a particle during zonal centrifugation runs in a given gradient solution.

In his derivation, μ_0 and ρ_0 are the light-end of viscosity and density, respectively. The coefficients λ_i' and ε_i' are characteristic constants for viscosity and density profiles in a rotor as a gradient solution. The reduced rotation time $A\tau$ is defined in terms of (a) radius of a rotor R, (b) angular velocity ω, (c) the light-end of viscosity and density μ_0 and ρ_0, (d) size of a particle dp, (e) density ratio between particle and the light-end of a gradient solution ρ_0/ρ_p, and (f) the centrifugation time t. Thus the reduced rotation time is given as

$$A\tau = \left[\frac{1}{Q} - 1\right]\frac{N}{18}\tau \tag{13}$$

in which

$$N = \left[\frac{\omega\rho_0 dp R}{\mu_0}\right]^2 \tag{13a}$$

$$\tau = \left[\frac{\mu_0 t}{\rho_0 R^2}\right] \tag{13b}$$

$$Q = \rho_0/\rho_p \tag{13c}$$

Quantitative evaluations of properties of the mathematical expressions derived were then made by using the following hypothetical gradient solution profiles:

$$\mu/\mu_0 = 1 + \lambda\zeta + \frac{\lambda^2}{2}\zeta^2 + \frac{\lambda^3}{6}\zeta^3 \tag{14a}$$

$$\rho/\rho_0 = 1 + \varepsilon\zeta + \frac{\varepsilon^2}{2}\zeta^2 + \frac{\lambda^3}{6}\zeta^3 \tag{14b}$$

where $\lambda_i = \lambda_i'/R$, $\varepsilon_i = \varepsilon_i'/R$, and $\zeta = r/R$. The characteristic constants for the gradient solution used were $\lambda = 0.5$, 1.0 $(0.5)^*$ 3.0, and $\varepsilon = 0.1$ $(0.1)^*$ 1.0. The numerical studies indicate the following conclusions.

(1) If the viscosity of a gradient solution decreases while the density of a gradient solution increases, a particle sediments much faster than in a medium of constant density and viscosity. Therefore, one may conclude that the viscosity of a gradient solution is the controlling factor in the decision of length of time required for a zonal centrifugation run for isopycnic banding and also for velocity sedimentation separation.

(2) The variation of the viscosity profile indicates that the steeper the profile, the longer is the time required for a particle to reach a given position.

(3) The centrifugation time for a particle to reach a certain position in a rotor is inversely proportional to the square of the particle diameter and to the square of the rotor speed.

(4) The particle diameter and the rotor speed are also inversely proportional to each other. If the diameter of a particle is one-tenth that of a reference particle, an increase in rotor speed to 10 times the original speed will result in the particle reaching the same position at the same time as the reference particle traveling at the original speed.

(5) The variation of the density profile also indicates that the steeper the profile, the longer is the time required for a particle to reach a given position.

(6) The variation in the density ratio between a particle and the light-end of a gradient solution shows the smaller the ratio, the shorter is the time for a particle to reach its isopycnic position.

The mathematical expressions were also tested from the zonal experimental data, numbers 813 and 859, at the Molecular Anatomy Program, Oak Ridge National Laboratory. The mathematical prediction of a particle's position in a rotor for a given rotation time agrees excellently with experiments for both runs. It is anticipated that these expressions will be eventually used in conducting a zonal centrifugation for design and predictions.

The Turnover Effect

When a particle-rich suspension is layered over a homogenous solution made with a freely diffusible solute, particulate material is observed to

---

*The number in parentheses indicates the interval for the increment.

stream to the bottom of the tube at normal gravitation force field at a velocity much greater than that exhibited by individual particles. This has been termed the "turnover effect" (*1*, *8*). With very large particles, statistical fluctuations in the particle concentration may play a minor role in producing these phenomena; however, as proposed by Brakke (*24*), differences in the diffusivity of the gradient solute and the sample particles appear to be the major cause. The effect is not prevented when the two-solution system is centrifuged at low speed. If a density gradient is employed, a marked decrease in this form of convective transport is seen (*25*). Since the shields used in high-speed swinging-tube centrifugation are opaque, the magnitude of the effect on published rate studies where no internal particle standard is used is not known. The effect has recently been examined by Schumaker (*11*) with no clear-cut decision as to the role of diffusion vs local density fluctuations. It is unfortunate that simple diffusion theory cannot be applied to the convective disturbances occurring at the interface between sample and gradient. The suggestion of Svensson, Hagdahl, and Lerner (*26*) that the sample be applied as a double gradient, positive with respect to the gradient solute (increasing in concentration in a downward or centripetal direction) and negative with respect to the sample, solves this problem at the beginning. However, as shown by Berman (*27*), an initially stable zone may become unstable as it moves down the gradient, either because the gradient changes slope and therefore capacity, or because the sedimenting zone becomes thinner at some point in the gradient and overloads it. Clearly the turnover effect is a serious problem which must be solved if effective use is to be made of rate-zonal centrifugation for preparative purposes.

There is no convincing evidence that the effect is not entirely due to differences in the diffusion coefficients of gradient solute and sample. It appears to involve three phases: (a) Diffusion which increases the density of the boundary zone in the sample layer and decreases the density in the gradient solution beneath. This double effect produces an instability.

(b) Fluid movement occurs when a sufficiently large volume of unstable fluid exists to be moved by gravitational force. To a first approximation the size of the moving droplets will be inversely proportional to the gravitational field. (c) As the droplets move they gradually disappear as their contents mix with the surrounding fluid. The rate at which this occurs is inversely proportional to some function of droplet size and the diffusivity of their contents. The net effect, given a sufficiently large centrifugal field, is to move the *minimum particle mass the minimum distance required to restore stability*. The turnover effect is observed in dynamically loaded and unloaded zonal centrifuges, and would have been readily seen with those rotors of the A-series which have transparent end-plates. Autocompensation of overloaded zones in zonal centrifuges due to movement of very small droplets a short distance is a key process and is the compelling reason for employing dynamic loading where high resolution and capacity are required. Autocompensation appears to occur at any level of the gradient where overloading may occur.

In addition to autocompensation, another process occurs concurrently in dynamically loaded rotors. Large sample zone particles which diffuse very slowly begin to sediment at a rate which may move them into the gradient faster than gradient solutes diffuse into the sample. Diffusion and sedimentation may thus partially compensate for each other.

A further stabilizing effect is provided by the heterogeneity of most biological suspensions. The narrow initial sample zone is rapidly spread out through the gradient, decreasing the sample load at the sample–gradient interface.

Gradient Capacity and Gradient Shape

As first pointed out by Svensson et al. (26), the theoretical capacity of a gradient is a function of its density slope. Adaptations of their gradient capacity equations (8, 24, 26–28) have been presented which suggest that the zones would be more Gaussian and less wedge-shaped. Experimentally, Brakke (29, 30) found that only a few percent of the theoretical load could be supported in swinging-bucket rotors. However, in a dynamically loaded zonal rotor, Spragg and Rankin (31) were able to demonstrate 60 to 70% of the theoretical capacity using shallow gradients.

It now appears that the problem of gradient capacity resolves itself into a very simple question; namely, what is the minimum gradient required to maintain gradient stability in the rotor? Whether the gradient is formed from sucrose, virus particles, subcellular components, or a mixture of these is apparently unimportant. This question remains to be answered experimentally for each class of rotor and condition of operation. From the data of Spragg and Rankin (31), the minimum gradient appears to be about 0.001 g/cc/cm for the B-XIV rotor. The value probably depends on rotor volume and configuration, temperature and temperature control, gradient viscosity, number of septa and the clearance between septa and the rotor wall, and the amount and frequency of vibrations originating in the drive system.

As noted by Berman (27), gradients can be constructed in which initially stable particle bands become unstable during sedimentation and broaden by autocompensation. The shape of gradients used for particle separation therefore deserves comment. A clear distinction should be made between the gradient as produced by the gradient engine (plotted as concentration vs volume) and the shape of the gradient as it exists in a spinning rotor (plotted as concentration vs radius). A gradient linear with respect to volume is concave when plotted as concentration vs radius because of the sector-shape of the zonal rotor compartments.

For the following reasons a gradient convex with respect to radius is desired: (a) greatest particle-supporting capacity is needed in the region of the sample; (b) as the difference between particle density and gradient density diminishes along the gradient, less gradient slope is required to support a given mass of particles; and (c) radial dilution in sector-shaped compartments decreases the particle concentration in a given zone, again requiring a gradient decreasing in slope with increasing radius.

Stability Analysis of Particle Sedimentation in Gradient Solutions

In studying the turnover effect, Sartory (16) has used a small perturbation analysis to determine whether the disturbance grows or decays in time for two layers of stationary diffusing solute in a common solvent in a gravitational field. His theory predicts that instability occurs under the much wider range of conditions $\bar{\rho}_2{}^0/\bar{\rho}_1{}^0 > (D_2/D_1)^{3/2}$ for sufficiently long times for thick upper layers, and under the conditions $\bar{\rho}_2{}^0/\bar{\rho}_1{}^0 > (D_2/D_1)^{5/2}$ for sufficiently long times for very thin upper layers. The quantities $\bar{\rho}_2{}^0$ and $\bar{\rho}_1{}^0$ are initial macromolecular (particle) density and initial gradient salt density, respectively. D_2 and D_1 are diffusivities of macromolecule in a gradient medium and gradient salt in the medium.

Recently, experiments have been conducted by Halsall and Schumaker (32) to determine the onset of turnover effect in diffusion experiments in zonal ultracentrifuge. They have found that the stability criteria for their experiment is $\bar{\rho}_2{}^0/\bar{\rho}_1{}^0 < (D_2/D_1)^{1.010}$. They concluded that their discrepancy from Sartory's prediction may be due to inhomogeneity or association–dissociation of the sample under study. Recently Meuwissen and Heirwegh (33) have shown experimentally that the stability of zones in liquid density gradient under a normal gravitational field depends on the strength of the supporting gradient.

Hsu (34) has made an analysis for stability of isopycnic banding in zonal centrifugation recently. Due to the isopycnic condition, the analysis can be made by the lumped parameter method and an analytical solution can be obtained. The result is summarized in the following.

A small pertubation analysis has been applied to the generalized Lamm sedimentation equation to determine whether the disturbance grows or decays in time for two stationary diffusing solutes [a gradient solute (1) and a macromolecule (2)] in a solvent (0) in a given centrifugal force field. The criterion for the stability has been obtained in terms of normal modes, so that the pertubations decay with time. It is found that the stability criterion is given by

$$\alpha r \leq \left\{ 1 - \frac{(D_{11} + D_{22})^2}{D_{11}D_{22} - D_{12}D_{21}} \left[\frac{D_{11} + D_{12}}{(D_{11} + D_{22})^2} \frac{\omega^2 d_2^2 r^2 \bar{\rho}_2}{18\eta} \left(3 - \frac{d \ln \eta}{d \ln r} \right) \right. \right.$$
$$\left. \left. + \frac{1}{4} \left(1 - \frac{1}{D_{11} + D_{22}} \frac{\omega^2 d_2^2 r^2 \bar{\rho}_2}{18\eta} \right) \right] \right\}^{\frac{1}{2}} \tag{15}$$

It is implied that the stability of an isopycnic banding is determined by the band width (αr), with the maximum band capacity as the right-hand of the equation.

The stability criterion given in the equation presents the unifying theory. The theory obtained from the previous observations during centrifugation, such as density inversion theory, and the theory from the infinitesimal pertubation analysis of diffusional phenomena under one gravitational field are inclusively represented.

It is interesting to note that Meuwissen and Heirwegh's conclusion (33), that stability depends on the shape or strength of the supporting gradient, is also represented by the terms of $[3 - (d \ln \eta)/(d \ln r)]$ and $\omega^2 d_2^2 r^2 \bar{\rho}_2/18\eta$ in Eq. (15). If the gradient increases with the cubic of the radius, the term $[3 - (d \ln \eta)/(d \ln r)]$ drops out. The shape of the gradient is not a factor for the stability. If the slope of a gradient viscosity is $(d \ln \eta)/(d \ln r) < 3$, the contribution of that term to the stability is negative, thus the maximum load capacity is reduced. If $(d \ln \eta)/(d \ln r) > 3$, the steeper the gradient solution the more the stability of a system. In this case the band width increases with the slope of a gradient solution, thus the range of the stable region increases. The same conclusion can also be drawn for the density inversion theory. If the shape of a gradient is $(d \ln \eta)/(d \ln r) < 3$, the second term on the right-hand side of Eq. (15) is positive. Therefore, the following situation has occurred in the gradient, $\bar{\rho}_2 < \rho_{grad}[3 - (d \ln \eta)/ (d \ln r)]$, since at an isopycnic point $\bar{\rho}_2 = \bar{\rho}_{grad}$. Thus density inversion does take place in order to return to the physical stable situation. If

$(d \ln \eta)/(d \ln r) > 3$, i.e., $\bar{\rho}_2 > \bar{\rho}_{\text{grad}}[3 - (d \ln \eta)/(d \ln r)]$, the system remains unchanged with the region. If a system is under one gravitational field, i.e., the terms containing $\omega^2 d_2{}^2 r^2 \bar{\rho}_2 / 18\eta$ drop out, the stability criterion becomes

$$\alpha r \leq \left[1 - \frac{(D_{11} + D_{22})^2}{4(D_{11}D_{22} - D_{12}D_{21})} \right]^{1/2} \tag{16}$$

The criterion is similar to that obtained by Sartory (16) except for the powers on the diffusivities. The four diffusivities for three-component systems can be obtained by Fujita and Gosting's procedure (38).

An additional characteristic in addition to previous theories, which appears in Eq. (15), is that one would like to increase the right-hand side of Eq. (15) so that the stability range can be increased. In order to increase the right-hand side of Eq. (15) for a given system, one will set the term

$$\frac{\omega^2 d_2{}^2 r^2 \bar{\rho}_2}{(D_{11} + D_{22})\eta} < 1$$

After rearranging, this becomes

$$\omega < \frac{1}{d_2 r} \left[\frac{(D_{11} + D_{22})\eta}{\bar{\rho}_2} \right]^{1/2} \tag{17}$$

in which r is the position where ideal isopycnic banding takes place. Equation (17) shows that the increasing of angular velocity in a zonal run does not improve the resolution. The angular velocity has to be constrained by Eq. (17). The physical interpretation of this phenomena is that with too high an angular velocity, the banding solute penetrates too deeply into a gradient solution, therefore an instability due to a density inversion or lack of strength in a supporting gradient will take place. Therefore, the angular velocity also has to be constrained.

The stability criterion is such a complicated phenomena that one should consider all the means to increase the right-hand side of Eq. (15). Thus Eq. (15) shows the unification of all previous theories.

Dispersion Coefficient of Sample in Gradient Solutions

The use of zonal centrifuges has two primary objectives: (a) separation and purification of biological materials from sample mixtures, and (b) concentration of separated biological material from a dilute solution. These two objectives are related. An optimum separation and purification is achieved by having a high resolution in each separated zone or band, and concentration increases as density of separated biological material in a band increases, likewise to achieve a high resolution in a separated zone. A band broadening effect in zonal centrifugation is due to a complicated phenomena of interaction of diffusion and sedimentation. In our investigation (36) we lumped all the contributing factors together and called it dispersion coefficients. In the determination of parameters which affect dispersion between sedimenting macromolecules and the sustaining gradient solution, dispersion coefficients of two sizes of polystyrene latex beads (diameters 0.91 ± 0.0058 and 0.312 ± 0.0022 μ) and bovine serum albumin (BSA) in sucrose and Ficoll (polysucrose by Pharmacia Fine Chemicals, Sweden), 9.5 to 10% w/v step-gradient and 10 to 25% w/v linear-with-volume gradient solutions, have been measured in an Oak Ridge B-XV zonal centrifuge rotor at 20°C. Dispersion coefficients were obtained from experimental data on intensity of UV absorbance of macromolecule vs volume fractions using the moment analysis technique.

In order to make the experimental results general and more useful, dispersion coefficients were correlated in a dimensionless number, the Schmidt number ($Sc = \mu_0/\rho_0 D$) as a function of various dimensionless parameters. Using results from step-gradient solution runs by a trial-and-error method, it is found that the following formula

$$Sc = 105[30F^{1.03}(Ta \cdot Q)^{-1} + 19Se^{0.19}\tau]^{1.02} \qquad (18)$$

fits all the data points within an average deviation of 2.5% and a maximum deviation of 6.3%. Various quantities in Eq. (18) are

$$Sc = \mu_0/\rho_0 D \qquad \text{Schmidt number}$$

$$F = \frac{\omega^2 t d_p^2 \rho_0}{\mu_0} \qquad \text{reduced centrifugal force field strength}$$

$$Ta = \frac{\omega d_p R \rho_0}{\mu_0} \qquad \text{Taylor number}$$

$$Q = \rho_0/\rho_p \qquad \text{density ratio}$$

$$Se = \frac{s\omega^2 R^2 \rho_0}{\mu_0} \qquad \text{reduced sedimentation coefficient}$$

$$\tau = \mu_0 t/\rho R^2 \qquad \text{reduced time}$$

The quantities μ_0 and ρ_0 are the viscosity and density of the gradient solution evaluated over the band-width values, respectively. D is the dispersion coefficient between the sample and a gradient solution, instead

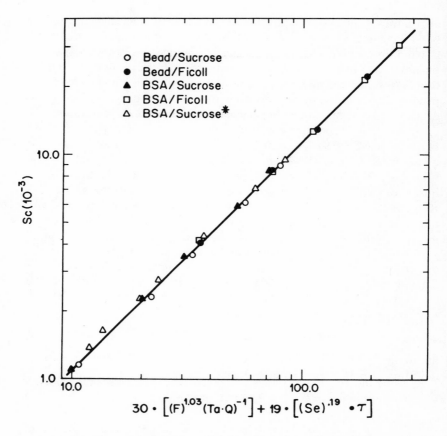

Fig. 3. A plot of Eq. (18) and general correlation of experimental data points.
(Data points △ were not used in the correlation.)

of the customary term of the diffusion coefficient. If $Q < 1$, a sedimentation takes place until $Q = 1$ (isopycnic point).

The results obtained from the linear-with-volume gradient solution runs were checked against Eq. (18). A good agreement exists. All the data points were within an average of 3.5% deviation and a maximum of 8.9% deviation. They are assumed to be within the range of experimental errors, so that a further correlation to improve the results was not made. Equation (18) and both correlations from a step-gradient solution and a linear-with-volume gradient solution are presented in Fig. 3.

Equation (18) and Fig. 3 may be used to estimate a dispersion coefficient on new biomaterials for which sizes of particles and their sedimentation coefficients are available.

Optimal Gradient Solution for Velocity Sedimentation

For the velocity-sedimentation method, a centrifugal field of specified force is chosen for the system. Because the rotor is operated at high speeds, the solute particles are forced to settle at appreciable rates toward the wall of a rotor. Thus it is basically a transport method. As pointed out previously (23), the viscosity of a gradient solution is the major factor in controlling the rate of particle movement. The design of an optimal density gradient solution is of importance in the velocity-sedimentation method.

Recently Hsu (37) has obtained analytically the optimal profile of a density gradient solution characterized by viscosity and density profiles which are expressed as functions of radial distance of a zonal rotor, the estimation of an optimal centrifugation time, so that the resolution of separation by the velocity-sedimentation method will be the maximum, and also the prediction of each particle's position in a rotor when centrifugation is stopped, so that the separated particles can be collected adequately. Therefore, the analysis will permit the rational selection of optimal operating conditions for mass separation of new biological materials by the velocity-sedimentation method.

By using the maximum principle, a performance index for the velocity-sedimentation for Particle 1 and Particle 2 is defined by (37):

$$\max\left[\Delta R = \int_0^{t_f}\left(\frac{dr_1}{dt} - \frac{dr_2}{dt}\right)dt\right] \tag{19}$$

The problem is to find a density gradient solution characterized by viscosity $\eta(r)$ and density $\rho(r)$ which maximize the difference of instantaneous position of two particles in a rotor so that the resolution of a separation will be maximum. The constrain conditions associated with the performance index equation, Eq. (19), are the Lamm's sedimentation equations for Particle 1 and Particle 2 together with the following

$$\frac{dr_1}{dt} = \frac{k_1\omega^2 r}{\eta(r)}\frac{\rho_1 - \rho(r)}{\rho_1 - \rho(H_2O, 20°C)} \tag{20a}$$

$$\frac{dr_2}{dt} = \frac{k_2\omega^2 r}{\eta(r)}\frac{\rho_2 - \rho(r)}{\rho_2 - \rho(H_2O, 20°C)} \tag{20b}$$

Equations (20a) and (20b) are obtained by combining the definition of sedimentation coefficient by the Svedburg and the usual way of converting the instantaneous sedimentation coefficient to the standard sedimentation coefficient in a 20°C water medium, Eq. (2). The quantity K_i is given as

$$k_i = s_i^0(H_2O, 20°C)\eta(H_2O, 20°C).$$

with the boundary conditions that the properties of the light-end density gradient solution at the rotor core r_c are

$$\rho(r_c) = \rho_0 \tag{21a}$$

$$\eta(r_c) = \eta_0 \tag{21b}$$

The optimal viscosity profile obtained is

$$\eta(r) = \eta_0\left(\frac{r}{r_c}\right)^{4/3} \tag{22a}$$

The optimal density profile is obtained by using the boundary condition, Eq. (21a), together with Eq. (22a), which gives

$$\rho(r) = (\rho_0 - \alpha)\left(\frac{r}{r_c}\right)^{2/3} + \alpha \tag{22b}$$

in which

$$\alpha = \frac{1 - \left(\frac{\rho_2}{\rho_1}\right)\left(\frac{s_2^0}{s_1^0}\right)\left(\frac{\rho_1 - \rho_H}{\rho_2 - \rho_H}\right)}{1 - \left(\frac{s_2^0}{s_1^0}\right)\left(\frac{\rho_1 - \rho_H}{\rho_2 - \rho_H}\right)} \tag{22c}$$

Equations (22) characterize the optimal density gradient solution for a velocity-sedimentation in terms of viscosity and density profiles.

Besides an optimal density gradient solution built in a rotor for a velocity-sedimentation, an important problem is that of when and where the maximum separation will take place, so that one can stop the centrifugation run and collect the separated fractions from a rotor either by draining the gradient out of the bottom of the rotor or by displacing it out of the top by using a heavier gradient solution. The final positions of Particles 1 and 2 are found to be

$$r_1(t_f) = r_c \left\{ \frac{\frac{4}{6}\left[\frac{s_1^0(\rho_1 - \alpha)}{\rho_1 - \rho_H} - \frac{s_2^0(\rho_2 - \alpha)}{\rho_2 - \rho_H}\right] + \frac{s_1^0(\rho_2 - \rho_1)}{\rho_1 - \rho_H}}{\left[\frac{k_1}{\rho_1 - \rho_H} - \frac{k_2}{\rho_2 - \rho_H}\right](\rho_0 - \alpha)} \right\}^{3/2} \tag{23a}$$

$$r_2(t_f) = r_c \left\{ \frac{-\frac{4}{6}\left[\dfrac{s_1^0(\rho_1 - \alpha)}{\rho_1 - \rho_H} - \dfrac{s_2^0(\rho_2 - \alpha)}{\rho_2 - \rho_H}\right] + \dfrac{s_1^0(\rho_2 - \rho_1)}{\rho_2 - \rho_H}}{\left[\dfrac{s_1^0}{\rho_1 - \rho_H} - \dfrac{s_2^0}{\rho_2 - \rho_H}\right](\rho_0 - \alpha)} \right\}^{3/2} \tag{23b}$$

The maximum separation ΔR thus obtained is

$$(\Delta R)_{max} = r_c \left\{ \frac{\frac{4}{3}\left[\dfrac{s_1^0(\rho_1 - \alpha)}{\rho_1 - \rho_H} - \dfrac{s_2^0(\rho_2 - \alpha)}{\rho_2 - \rho_H}\right]}{\left(\dfrac{s_1^0}{\rho_1 - \rho_H} - \dfrac{s_2^0}{\rho_2 - \rho_H}\right)(\rho_0 - \alpha)} + \frac{(\rho_2 - \rho_1)}{(\rho_0 - \alpha)} \right\}^{3/2} \tag{23c}$$

The optimum centrifugation time thus obtained is

$$\tau_f = \frac{1}{AB} \sum_{n=0}^{\infty} \frac{3M^n}{2(n+2)} \zeta_c^{[2(n+2)/3]}$$

$$\times \left[\left\{ \frac{\frac{4}{6}\left[\dfrac{k_1(\rho_1 - \alpha)}{\rho_1 - \rho_H} - \dfrac{k_2(\rho_2 - \alpha)}{\rho_2 - \rho_H}\right] + \dfrac{k_1(\rho_2 - \rho_1)}{\rho_1 - \rho_H}}{\left(\dfrac{k_1}{\rho_1 - \rho_H} - \dfrac{k_2}{\rho_2 - \rho_H}\right)(\rho_0 - \alpha)} \right\}^{n+2} - 1 \right]$$

$$\tag{24}$$

Equation (24) implies that the length of centrifugation for a velocity-sedimentation is determined by this formula in which the density gradient solution is characterized by Eq. (22).

These results are based on the assumption that the particles are still far away from their respective isopycnic points, so that sedimentation is dominant in the mutual interference of diffusion and sedimentation. Hence changes of a density gradient solution profile due to diffusion are negligible. In a velocity-sedimentation, the centrifugation time is generally very short. Therefore, concentrations of macromolecules and a density gradient solution do not disperse appreciably by molecular diffusion. Also, the partial specific volumes of each component are assumed constant. This implies that the solution is regular. The solution may not be regular but can be corrected by an activity coefficient. These assumptions are considered to be reasonable, and they permit the development of an analytical solution with relative ease.

The estimates of an optimal centrifugation time and the positions of particles serve as a guide for designing velocity-sedimentation runs. It is perhaps very difficult to build an optimal density gradient solution as specified in Eq. (22) by a single gradient solute. A mixed solutes gradient solution probably can provide the viscosity and density profiles as designed.

INTRODUCTION OF NEWER TECHNIQUES (5)

The sedimentation rate of a particle, its banding density, or both of these properties may be altered in a density gradient in a manner useful for achieving separations. These factors involved are:

(1) Precipitation (increase in effective particle size) produced by diffusion of a precipitant into a sample zone.

(2) Resolubilization in a gradient negative with respect to the precipitating agent.

(3) Alteration in volume and density by change in osmotic pressure or biochemical environment.

(4) Specific binding to ions or substances, altering sedimentation rate or banding density.

(5) Sequential dissection by sedimentation through zones of immobilized reagents.

These techniques, singly or in combination, open up many new interesting avenues of research.

Precipitation by a Diffusion Precipitant

Soluble sample materials of high or of low molecular weight may be precipitated by the diffusion of a suitable precipitant into the sample zone. Inorganic salts in aqueous solution may be caused to precipitate, often in the form of small crystals, as an organic solvent miscible with water (and in which the salts are not soluble) is allowed to diffuse in. Similarly, organic substances insoluble in water may be precipitated out of alcohol or acetone by water diffusing into the sample zone. High-molecular-weight substances such as proteins and nucleic acids may be precipitated by water-miscible organic solvents or heavy metal salts. The precipitating agent may be in either the overlay, in the underlying gradient, or in both. In all instances care must be taken to ensure that the proper density increments are present to ensure stability from a gravitational viewpoint. When ethyl alcohol is used as a solvent, its density may be adjusted with 2-chloroethanol, for example.

The unique feature about precipitation in a centrifugal field is that precipitated particles sediment out of the sample zone as they increase in size. Since precipitation occurs in response to a diffusion gradient, different substances may precipitate at different times and levels in the diffusion gradient, tending to minimize cross-contamination in the particles formed. Particle size will be a function of the centrifugal field used, and with reasonably high forces only very tiny particles will be formed. If the particles are not soluble at any level as they sediment through the gradient, they may be fractionated by isopycnic banding. This method of preparative particle separation has not been widely applied to precipitates although the crystal densities of a large number of substances are known.

Gradient Resolubilization

As precipitated particles are sedimented through a gradient, the composition of the gradient may vary in such a way that the particles go back into solution (4, 41, 42). One of the advantages of this method is that the time during which a substance is precipitated is rather short, especially where a high centrifugal field is employed. The rate at which a sedimenting particle will dissolve increases as the size of the particle increases. Interest in these processes centers chiefly around the possibility of adapting them to continuous on-stream protein fractionation.

Alteration in Volume and Density by Change in Osmotic Pressure or Biochemical Environment

Gradients may be prepared having similar density slopes, but having osmotic pressure gradients which may vary, and in the case of specially prepared sucrose-dextran gradients, may approach zero (40). Where particles differ in their response to an osmotic pressure gradient, the gradient may be adjusted to maximize the separation. Mitochondria appear to shrink appreciably more than do peroxisomes in sucrose gradients. The mitochondria initially sediment ahead of the peroxisomes; however, the latter overtake them and band at a slightly denser level. For optimal rate separations it appears advantageous not to employ isotonic gradients, but rather to let the mitochondria shrink and sediment more rapidly ahead of the lysosomes. Sedimentation rates may also be affected by the biochemical environment quite apart from osmotic effects. The volume changes produced in isolated nuclei by small amounts of divalent cations (41, 42) and in mitochondrial volume by ATP or thyroxine are examples of this.

Specific Binding to Ions or Substances Altering Sedimentation Rate or Banding Density

The banding density of smooth and rough endoplasmic reticulum fragments, and subfamilies of these, may be altered by small changes in the ionic environment (40). The use of uranium or ferritin-labeled antibodies to specifically alter the banding density of an organelle or fragment does not appear to have been explored. A simple and frequently used method for increasing the sedimentability of haemagglutinating viruses is to allow them to attach to red cells which are easily sedimented. As the virus desorb, they are easily separated from the red cells by centrifugation (43).

Sequential Dissection by Sedimentation through Zones of Immobilized Reagents

Reagents having low sedimentation coefficients may be incorporated at different levels in a density gradient to attack larger particles moving through the gradient. At the MAN Program the method was employed in early studies on the extraction of histones from nuclei sedimenting through gradients of increasing acidity, and more recently detergent gradients have been used to solubilize microsomes. As more interest develops in the dissection of membranes and ribosomes, these methods will probably be more widely employed.

DISCUSSION

Preparative zonal centrifugation is now in a state of very rapid growth, with new rotor systems and separative techniques appearing at frequent intervals. The evaluation of these methods and comparison with other methods in a variety of specific instances will require a period of several years.

We conclude the rotor development should proceed in those directions:

(1) The limits of speed, capacity, volume, and resolution should continue to be explored by using systems which exhaust presently available technology.

(2) The improvement of zonal centrifugation should be directed to an improvement in loading and unloading methods to reduce the dispersion from these operations. Therefore, the separated resolution can be adequately maintained during the unloading.

(3) An optimal design for rotor configuration is required for a par-

ticular separation. Transport phenomena of particles in a rotor with respect to various configurations has to be studied in detail so that various factors affecting separation in a rotor can be evaluated quantitatively.

Acknowledgments

The author wishes to thank Dr. N. G. Anderson for his constant encouragement and a NSF Grant GK-38341 for support during this work.

REFERENCES

1. N. G. Anderson, "Studies on Isolated Cell Components VIII, High Resolution Gradient Differential Centrifugation," *Exp. Cell. Res.*, 9, 446 (1955).
2. N. G. Anderson, "Zonal Centrifuges and Other Separation Systems," *Science*, 154, 103 (1966).
3. N. G. Anderson, ed., *National Cancer Institute Monograph 21*, U.S. Government Printing Office, Washington, D.C., 1966.
4. N. G. Anderson, "An Introduction to Particle Separations in Zonal Centrifuges," in *National Cancer Institute Monograph 21*, U.S. Government Printing Office, Washington, D.C., 1966, p. 9
5. N. G. Anderson, "Preparative Particle Separation in Density Gradients," *Quart. Rev. Biophys.*, 1(3), 217–263 (1968).
6. N. G. Anderson, H. P. Barringer, J. M. Amburgey, Jr., G. B. Cline, C. E. Nunley, and A. S. Berman, "Continuous-Flow Centrifugation Combined with Isopycnic Banding: Rotors B–VIII and B–XI," in *National Cancer Institute Monograph 21*, U.S. Government Printing Office, Washington, D.C., 1966, p. 199.
7. H. P. Barringer, N. G. Anderson, and C. E. Nunley, "Design of the B–V Continuous-Flow Centrifuge System," in *National Cancer Institute Monograph 21*, U.S. Government Printing Office, Washington, D.C., 1966, p. 191.
8. M. K. Brakke, "Zone Electrophoresis of Dyes, Proteins, and Viruses in Density Gradient Columns of Sucrose Solutions," *Arch. Biochem. Biophys.*, 55, 175 (1955).
9. M. K. Brakke, "Nonideal Sedimentation and the Capacity of Sucrose Gradient Columns for Virus in Density-Gradient Centrifugation," *Ibid.*, 107, 188–403 (1964).
10. M. K. Brakke, "Density-Gradient Centrifugation: Nonideal Sedimentation and the Interaction of Major and Minor Components," *Science*, 148, 387–389 (1965).
11. V. N. Schumaker, "Zonal Centrifugation," in *Advances in Biological and Medical Physics* (C. A. Tobias and J. H. Lawrence, eds.), Academic, New York, 1967, pp. 245–339.
12. T. Svedberg, *Z. Phys. Chem.*, 127, 51 (1927).
13. T. Svedberg and K. O. Pederson, *The Ultracentrifuge*, Clarendon Press, Oxford, 1940.
14. J. W. Williams, *Ultracentrifugal Analysis in Theory and Experiment*, Academic, New York, 1963.
15. S. R. de Groot and P. Mazur, *Non-Equilibrium Thermodynamics*, 2nd ed., North-Holland, New York, 1962.
16. W. K. Sartory, "Instability in Diffusing Fluid Layers," *Biopolymers*, 7, 251–263 (1969).

17. S. P. Spragg and C. T. Rankin, Jr., *Biophys. Acta*, *141*, 164 (1967).
18. N. G. Anderson, *Semiannual Report of the U.S. AEC*, ORNL-1953, August 15, 1954, p. 17.
19. H. Elrod, N. G. Anderson, L. C. Patrick, and J. C. Shinpaugh, *J. Tenn. Acad. Sci.*, (1968).
20. H. W. Hsu and N. G. Anderson, *Biophys. J.*, *9*, 173–188 (1969).
21. H. W. Hsu, *Separ. Sci.*, *8*, 537 (1973).
22. H. D. Pham and H. W. Hsu, *Ind. Eng. Chem., Process Des. Develop.*, *11*, 556 (1972).
23. H. W. Hsu, *Separ. Sci.*, *6*, 699–714 (1971).
24. M. K. Brakke, *Adv. Virus Res.*, *7*, 193–224 (1960).
25. N. G. Anderson, "Techniques for the Mass Isolation of Cellular Components," in *Physical Techniques in Biological Research*, Vol. 3, *Cells and Tissues*, (G. Oster and A. W. Polister eds.), Academic, New York, 1956, pp. 299–352.
26. H. Svensson, L. Hagdahl, and K. D. Lerner, *Sci. Tools*, *4*, 1–10 (1957).
27. A. S. Berman, "Theory of Centrifugation: Miscellaneous Studies," in *National Cancer Institute Monograph 21*, U.S. Government Printing Office, Washington, D.C., 1966, p. 41.
28. M. Gehetia and E. Katchalski, *J. Chem. Phys.*, *30*, 1334–1339 (1959).
29. M. K. Brakke, *Archs. Biochem. Biophys.*, *107*, 388–403 (1964).
30. M. K. Brakke, *Science*, *148*, 387–389 (1965).
31. S. P. Spragg and C. T. Rankin, Jr., *Biochem. Biophys. Acta*, *141*, 164–173 (1967).
32. H. B. Halsall and V. N. Schumaker, *Biochem. Biophys. Res. Commun.*, *43*, 601–606 (1971).
33. J. A. T. P. Meuwissen and K. P. N. Heirwegh, *Ibid.*, *41*, 675 (1970).
34. H. W. Hsu, *Math. Biosci.*, *13*, 361 (1972).
35. H. Fujita and L. J. Gostings, *J. Phys. Chem.*, *64*, 1256 (1960).
36. R. K. Genung and H. W. Hsu, *Separ. Sci.*, *7*, 249 (1972).
37. H. W. Hsu, *Math. Biosci.*, *23*, 179 (1975).
38. N. G. Anderson, C. A. Price, W. D. Fisher, R. E. Canning, and C. L. Burger, *Ann. Biochem.*, *7*, 1–9 (1964).
39. N. G. Anderson, H. P. Barringer, E. F. Babelay, and W. D. Fisher, *Life Sci.*, *3*, 667–671 (1964).
40. G. Dallner and R. Nilsson, *J. Cell. Biol.*, *31*, 181–193 (1960).
41. J. F. Albright and N. G. Anderson, *Exp. Cell Res.*, *15*, 271–281 (1958).
42. N. G. Anderson and K. M. Wilbur, *J. Gen. Physiol.*, *35*, 781–797 (1952).
43. F. Hoyle, B. Jolles, and R. G. Mitchell, *J. Hyg.*, *52*, 119–127 (1954).

AUTHOR INDEX

Numbers in parentheses are reference numbers and indicate that an author's work is referred to although his name is not cited in the text. Italic numbers give the pages on which the complete references are listed.

SUBJECT INDEX

Date Due

Due	Returned	Due	Returned
MAY 03 '78	MAY 8 '78		
JUL 3 '78	JUL 7 '78		
DEC 16 '78	OCT 13 '78		
JUN 9 1981	JUN 9 '81		
OCT 2 2 1981	OCT 23 '81		
NOV 1 5 1981	NOV 8 '81		
JUN 0 2 1982	JUN 0 2 1982		
FEB 0 7 1984	FEB 0 8 1984		
APR 0 1 1991	JUL 2 3 1991		
JUN 0 9 1995	SEP 0 4 1995		